SILK & VENOM

JAMES O'HANLON has travelled around Australia and the globe uncovering the secret lives of insects and spiders. If it is small, mysterious and lacks a backbone, James has an insatiable desire to find out what it is and what it does. He has published more than 30 academic papers and his popular science writing has appeared in ABC News, *Australian Geographic*, *The Conversation* and *Biosphere Magazine*. He is an award-winning science communicator and was the 2021 recipient of the Varuna–New England Writers' Centre fellowship.

SILK & VENOM

The incredible lives of spiders

JAMES O'HANLON

NEWSOUTH

UNSW Press acknowledges the Bedegal people, the Traditional Owners of the unceded territory on which the Randwick and Kensington campuses of UNSW are situated, and recognises their continuing connection to Country and culture. We pay our respects to their Elders past and present.

A NewSouth book

Published by
NewSouth Publishing
University of New South Wales Press Ltd
University of New South Wales
Sydney NSW 2052
AUSTRALIA
https://unsw.press/

First published 2023

10 9 8 7 6 5 4 3 2 1

NATIONAL LIBRARY OF AUSTRALIA

A catalogue record for this book is available from the National Library of Australia

ISBN 9781742237817 (paperback)
9781742238838 (ebook)
9781742239774 (ePDF)

Internal design Josephine Pajor-Markus
Cover design Regine Abos
Cover image Golden orb-web spider by Mathew Schwartz/Unsplash
Printer Griffin Press

This book is printed on paper using fibre supplied from plantation or sustainably managed forests.

UNSW
SYDNEY

CONTENTS

PROLOGUE

Picture the web of a common garden spider, a rough circle strung in the space between two plants. Spiralling strands of silk are draped over tougher radial threads. Small droplets of a glue-like substance are dotted along the silk strands. The web is delicate and fierce, soft enough to billow in the breeze yet strong enough to ensnare flying prey.

In the centre of this little masterpiece sits its owner, a small orb-web spider. Her web is both a sophisticated trap and a sensory experience. Her eyesight is poor so she must use the web to listen to the world around her, encoded in the gentle vibrations of silk struck like miniature piano strings. With legs poised against radial threads she hears the crescendo of a gust of wind and the staccato of leaves tapping against her silk threads as they fall from above. But these vibrations she ignores as she waits for the tell-tale tremors of an insect ensnared in her trap. She is so attuned to the subtle tones of her web that she can deduce what type of prey has landed. Buzzing wings will tell her that a fly is ensnared, whereas violent shudders signal something larger, like a grasshopper kicking to free itself of the sticky tendrils. The web is her conduit to the world; it is her ears and eyes, an extension of herself.

Finally, she hears something – a quick strum and then silence. A few seconds later she hears the same sound again, the familiar sound of a prospective meal emanating from the outer edge of her web. It is neither a fly nor a grasshopper. It is something new, something different. Unsure of what this new meal is, and unable to see from this distance, she approaches slowly. She treads cautiously across her network of silk, triangulating on the sound. When she arrives at the edge of the web, all is quiet. There is nothing to be found.

A sharp tube is thrust into the back of her head, piercing her exoskeleton. She struggles to move but is impaled in place. Fluids are pumped through the tube into her body cavity and her insides begin to slowly dissolve. Her struggles of resistance slow and come to a stop as her innards are gradually liquefied.

There had been no prey for the spider to find because standing above her was an assassin bug. *Stenolemus bituberus* is a master of stealth that hunts and feeds on orb-web spiders. They are frail looking creatures with angular appendages almost as delicate as the silk on which they stand. Their entire morphology is specialised for sneaking across the sensitive threads of spiders' webs. To capture a spider, they must step into a loaded trap. As they stalk across the taut web, any vibrations they create can give away their presence, either scaring the spider away or turning the tables on themselves. When they come up against a strand of silk in their path, assassin bugs use their sharp mouthparts to slice through it and continue onward. Should a gust of wind shudder the web, they opportunistically advance a few steps forward while the sound of their footfall is muffled.

Stalking is only one of the techniques in the assassin bug's arsenal. Using its front pair of legs, the assassin bug plucks the spider silk, creating sounds that resemble an insect ensnared in the web. It is a risky strategy that, if used correctly, will lure the spider close. As the spider nears, the assassin bug will raise its head high and ready its long, piercing proboscis. Once the spider is in range, the assassin bug pounces, impaling the spider from above. Digestive fluids are pumped down the bug's tube-like mouthparts and the spider's insides are broken down. Once the spider's tissues have been reduced to a soup, the assassin bug draws the fluid in. And the spider, the fabled predator with her sophisticated trap, is left as an empty husk dangling from the web.

—

I have spent much of my adult life surrounded by people who adore spiders – scientists and natural historians who have devoted their lives to understanding the variety, beauty and hidden secrets of these eight-legged wonders. The spiders they tell stories about are not evil denizens of haunted houses or monstrous creatures of fantasy. They are caring parents, and social family members; problem solvers and cunning masters of disguise. There are tales of swimming spiders that live underwater inside silk-lined bubbles, and miniature aviators that glide over oceans on silken kites. Spiders are, without a doubt, incredible.

But this perspective is alien to many people. Somewhere in our history, we have learnt to fear and vilify spiders, and they have become associated with deception, death and decay. I often wonder how the humble orb-web spider – small, slender, and prey to so many other animals – can incite genuine fear. Admittedly, there are a small number of spiders that can harm humans, but then again so can many other animals. Even now, as I sit writing this, I have a large dog sitting on the floor beside me. His name is Wallace, he weighs over 50 kilograms and has the bite force to tear through my flesh and crush the bones in my feet, a possibility that is statistically much more likely than a harmful spider bite.

We have formed a relationship with dogs and other animals that allows us to forgive their trespasses and embrace their variety. The relationship we have formed with spiders is different and largely unjustified. It seems many people will flat-out refuse to believe that a spider can be anything other than the preconceived villain that has been installed in their psyche. This needs to change. I hope by sharing stories about the amazing lives of spiders that we can start to form a better relationship with them. Who knows, perhaps one day your squeals upon encountering a spider will become squeals of delight. If nothing else, then perhaps you can have a little sympathy for those poor orb-web spiders out there being digested alive in their own skins.

Chapter 1

WHY DON'T WE LIKE SPIDERS?

A few years ago, I went to the opening of a new exhibit all about spiders at a museum in Sydney. It was unlike any exhibit I had seen before. There were live animals on display, arcade-style games where people could play the role of a spider catching its prey, larger-than-life models that gave an up-close look inside spiders' bodies, and live demonstrations of spider venom milking. The fun began as soon as I walked inside. In the entrance to the exhibit, projected onto the floor in front of me, was an interactive video game. The ground was a lush forest floor of orange and green leaves. When I looked closer there were hundreds of tiny animated spiders walking around the leaf litter. As I stepped across the projected image the spiders quickly darted to the side, avoiding my footsteps. It was a beautiful combination of art, play and technology that made the simple act of entering an exhibit joyous.

A few days later I was speaking to a friend who worked at the museum. I mentioned how much I loved the exhibit and how that display had me hooked from the moment I walked in. My friend gave a half-smile and told me that the museum had to move the animated spider display from the entrance after only a single day of the exhibit being opened. Eager museum guests found themselves paralysed in fear, unable to go inside, because of the fake spiders blocking their path. The display was demoted to a back corner where it could be easily avoided, and soon became a parking lot for discarded prams and backpacks.

I have no idea what a spider phobia must feel like. Rather, I have made a living out of getting up close and personal with all kinds of creepy-crawlies. As a scientist I have spent years collecting and studying spiders, wasps, ants, bees, praying mantises, stick insects, grasshoppers, crayfish, frogs, lizards, the list goes on. For me, there's something fascinating about the little things that scurry around our feet. They are all around us and live incredible and complex lives that go largely unnoticed. Being a scientist is like lifting a veil off a hidden miniature world to discover its secrets.

Whenever I tell people that I am a zoologist and that I study animal behaviour, they immediately assume I must study fluffy things like koalas or kangaroos. The fact that insects, spiders and other backbone-lacking critters are even considered animals in the first place is often forgotten, even though they make up over 99 per cent of all animals that we know of. When I start talking about spiders, some will smile politely as they eye off nearby exits, while others make their disgust fantastically clear by cringing and moaning. While I am fully aware there are a very small number of spiders that have the potential to be harmful, it doesn't affect my daily life or stop me from doing my job. And if being scared of live spiders is an alien concept to me, then the thought of being scared simply by an image of a spider, like the ones in the museum display, is even more incomprehensible.

How we feel about spiders

There are over 50 000 species of spiders and of these thousands of species, only a very small number are capable of causing harm to people. In most cases spider bites are small 'pin-pricks' that cause local swelling and don't require any medical attention at all. Regardless, fear of spiders is one of the most commonly reported phobias along with things like snakes, heights and flying. Scientists have attempted to understand why people are afraid of spiders by surveying self-identified arachnophobes about what they don't like about spiders. The most common responses are broad statements about their 'legginess' and the 'way they move'. Surprisingly, spider fangs and venom hardly ever come up as a reason for being afraid of them. Most people fear spiders simply because of the way spiders make them feel, as opposed to a rational assessment of their potential danger. Spider phobias make even less sense when we compare them to something like scorpion phobias. Admittedly, scorpions are not everyone's favourite household pet, but clinically significant scorpion phobias are rare. This is strange given that scorpions are very similar to spiders; they are arachnids, they are small and do spidery things like crawl around on the ground and use venom to catch their prey, and are sometimes venomous to humans. Some data suggest more people die from scorpion stings every year than from spider bites, and the effects of scorpion stings are arguably much worse than those from spider bites. A common symptom of scorpion stings is something called cholinergic crisis, colloquially known as SLUDGE syndrome; which, to put it bluntly, is every possible fluid pouring out of every possible orifice in your body. Not to mention searing pain, and the risk of seizures, strokes and heart attacks. The take-home message here is that spider phobias are not rational. By definition, all phobias are irrational. If butterflies, for example, were to pose a genuine threat to your safety, then we wouldn't call fear of butterflies a phobia and give it a fancy name like 'lepidopterophobia', we would just call it common sense.

In addition to genuine phobias, there is a general public disdain for spiders that is largely unjustified and hard to rationalise. The cultural norm is to dismiss spiders as 'icky' things to be purged from our homes. In 2013 the journal *American Entomologist* conducted a survey to understand why, even in the entomological community, there are people with strong spider aversions. Entomologists, otherwise known as insect researchers, spend their lives passionately studying small crawly *six-legged* things. But for some entomologists, small *eight-legged* things are too hard to handle. Strangely, the fact that spiders have a lot of legs was reported as a common factor in entomologists' spider aversions. One survey respondent even identified as a forensic entomologist, which is a fancy term for someone who picks maggots out of rotting corpses for a living. And even *they* admitted spiders gave them the heebie-jeebies. Overall, the survey respondents were aware that their spider fears were unjustified, but this knowledge still didn't stop them from feeling the way they did.

If I had to guess why spiders' bad reputations persist, I would argue that a lot of it has to do with the bad habits we have formed. A few years back I was teaching a university zoology class where, in each laboratory session, students inspected and dissected real animal specimens. We cut out fish guts to see what they had been eating, we scoured through museum collections in search of treasured horseshoe crabs and

Guinea worms, we even went exploring through the insides of a garden snail and discovered their amazing genitalia (seriously, snail junk is crazy, but that's a story for another book). The students embraced each challenge that was thrown at them. Until the day I brought out some live golden orb-web spiders – gorgeous and harmless spiders with shimmering black and gold legs. Suddenly, what seemed like all of the women in the class squealed in unison and began squirming in their seats. I never expected that this keen group of adventurers and future scientists would respond like this, and I especially didn't expect them to conform to such dull gender stereotypes.

In hindsight, I don't think that the students were actually disgusted by the spiders. Instead, I think they were simply behaving how they thought they were supposed to behave around spiders. When faced with the topic of spiders, in the absence of any other stories to tell, we default to the rehearsed and socially acceptable responses of 'eww', or 'yuk' and move on. This is why, I think, we need better spider stories on hand to replace the old clichéd tropes of spiders being gross and creepy. This is also why, despite my disappointment in hearing that the museum removed the interactive spider game from the exhibit entrance, I think they made the right decision. By removing an initial barrier, the museum made it more accessible to visitors who might be afraid of spiders, hopefully allowing them the opportunity to explore at their own pace and within their own limits, learning new things and beginning to see spiders in a new light.

Spider surveys

To study how people feel about spiders, psychologists use self-assessment surveys where people can answer questions about their feelings towards spiders. Their responses are given a score based on how spider-aversive they are. There is no pass or fail grade as to whether someone is afraid of spiders or not, but the scores are useful for comparing between people or comparing how people feel about spiders before and after treatment for phobias. There are a few different spider questionnaires around, with names like the 'Spider Fear Questionnaire' or the 'Fear of Spiders Questionnaire'. I know that many people will be reading this book because they are already fans of spiders, but I hope there are more than a few of you reading this who are a little bit curious but still not quite sure about spiders. It would be interesting to see where you rank on a survey about spider attitudes. I assume that you are reading this book for the sheer fun of it, so I am not going to submit you to a formal psychological evaluation. Instead, I have made up my own survey with a less judgemental title – *The Spider Vibes Survey*. It's just for fun and is a lot simpler and shorter than the real surveys. It consists of ten statements where you can either agree or disagree along a scale of 1 to 5. At the end you can sum up your scores and see where you fit between the lowest score of 10 and the highest score of 50. Give it a go and see if any of the questions make you think about your current relationship with spiders. You could even test your friends and family and see how they feel about spiders compared to you.

THE SPIDER VIBES SURVEY

Q1 I think spider webs are yucky

Disagree 1 2 3 4 5 Agree

Q2 If I saw a spider now, I wouldn't want to get close to it

Disagree 1 2 3 4 5 Agree

Q3 The thought of a spider crawling on me makes me very uncomfortable

Disagree 1 2 3 4 5 Agree

Q4 If I saw a spider in the house, I would need someone else to remove it for me

Disagree 1 2 3 4 5 Agree

Q5 I don't want to be in the same room as a spider

Disagree 1 2 3 4 5 Agree

Q6 I worry about being bitten by spiders

Disagree 1 2 3 4 5 Agree

Q7 I think spiders are ugly

Disagree 1 2 3 4 5 Agree

Q8 I don't like the way spiders move

Disagree 1 2 3 4 5 Agree

Q9 I think most spiders are dangerous

Disagree 1 2 3 4 5 Agree

Q10 I don't like spiders

Disagree 1 2 3 4 5 Agree

The truth about spiders

On 6 April 2022, the scientific community celebrated the description of the 50 000th spider species known to science. The recipient of this honour was a small jumping spider from South America, now called *Guriurius minuano*. There are so many spider species, and new species being found so frequently, that spider taxonomists are apparently getting bored coming up with names for them. There's a spider named after Bernie Sanders (*Spintharus berniesandersi*), another after Angelina Jolie (*Aptostichus angelinajolieae*), two named after David Bowie (*Heteropoda davidbowie* and *Spintharus davidbowiei*), and there's even an entire genus of spiders named *Pinkfloydia*. Spiders live in every corner of every continent but Antarctica, and they have been around for at least 300 million years; the oldest fossil spider ever found is from the Carboniferous period – more than 50 million years before the first dinosaurs. Spiders range from the tiniest *Patu digua* from Colombia (which is less than half a millimetre wide and is a serious contender for the title of 'most adorable spider'), to the giant huntsman *Heteropoda maxima* from Laos (which gets up to about 30 centimetres wide and is beautiful in its own way). Their diversity is hard to imagine; there are ant-mimicking spiders, colour-changing crab spiders, Moroccan cartwheeling spiders, and pelican spiders that look like something went wrong during spider-yoga class, just to mention a few.

This book aims to help those of us who are a little 'arachno-challenged' to appreciate spiders by sharing stories about just a few of these amazing critters. I won't attempt to rationalise spider-positivity by dwelling on statistics about spider bites and biodiversity. If climate change and vaccinations have taught us anything, mere facts are not the powerful tools of persuasion that they should be. This is not a guide to curing spider phobias, nor is it a textbook on the finer details of spider biology. I will not dive into the physiology of their book lungs, or guide you through spider taxonomy, no matter how fascinating those topics may be. Nowhere in this book will you see the word 'opisthosoma'. Well, except for that one just there. I will not be presenting you with any spider phylogenetic trees, nor will I expect that you care what a phylogenetic tree is. Instead, I will tell stories. Then, hopefully, when the topic of spiders comes up at your next evening soiree, you will forego the usual rehearsed responses about spiders being creepy-crawlies, and instead offer some more interesting stories about, say, lactating spider-mothers and genetically engineered spider-goats.

Having said that, let's cover a few essential details. We all know that spiders are arachnids, but not all arachnids are spiders. Arachnida is a larger class of animals containing things like scorpions, ticks and mites, to name just a few. True spiders are a specific arachnid order called Araneae. There are many other things that have the name spider attached to them; camel spiders, sea spiders, spider crabs, spider cheese, spider monkeys, spider orchids, ice-cream spiders, so on and so forth. These are not true spiders, and some of them aren't even animals, so we won't be dealing with them here.

You probably already know that spiders have eight legs, and eight-ish eyes. They also have two main body segments: the head up the front, and the abdomen at the back (which is also sometimes called the 'O-word' that I said I wouldn't mention earlier). The head is where all the showy stuff happens, it's got the legs, eyes, mouth and fangs on it. The abdomen is where all the secret inside stuff happens. It's got all manner of cool organs hidden away inside a big round exoskeleton, like silk glands, book lungs, and stomach-ey bits.

Spiders are usually split up into two big groups: the families Mygalomorphae and Araneomorphae. The mygalomorphs are the big-hairy-scary looking ones like tarantulas, funnel-webs and trapdoor spiders. They are identified by having parallel fangs that point straight down from their head. The araneomorphs have fangs that point towards each other so that they can pinch things rather than push down into things. This group includes all of the other more delicate-looking spiders like orb-web spiders, jumping spiders, huntsman spiders, wolf spiders and crab spiders. There is actually a third group of spiders called the Mesothelae. They're also known as 'segmented spiders' because of the visible segments on their abdomens, which is seen as a very ancestral trait. There are only a small number of them (approximately 140 species, only found in Asia, which sounds like a lot but it's a drop in the ocean in terms of spider diversity), so most texts tend to overlook them a bit. And to be fair, so does this one. We really don't know much about them other than that they live on the ground and hide in burrows. So, this book sticks to the status quo of focusing on the two big groups, the mygalomorphs and the araneomorphs. And while the big-bitey-hairy mygalomorphs might be the first thing that comes to people's minds when they think about spiders, they are actually in the minority and only make up less than ten per cent of all spider species. The remaining approximately 90 per cent belong to the araneomorphs. Most spiders that you come across won't have the big piercing fangs you've seen in horror movies; they actually have small pinching mouthparts. Araneomorphs are considered the most 'modern' group because their features evolved later in spiders' evolutionary history. The mygalomorphs are more ancient, while the Mesothelae are more ancient still, and are most similar to what the first spiders to walk this Earth would have looked like.

Let's tell better spider stories

If you are sceptical as to whether storytelling has the power to change how society views misunderstood animals, we only need look at how the discovery of whale song changed the world. It's hard to believe now, but before the 1960s nobody cared about whales all that much, and those people who did were mostly interested in killing them. Whales were hunted globally for their meat, blubber and oil, and by the 1960s many species were nearing extinction. In popular culture they were portrayed as terrible beasts, like the limb-devouring Moby Dick or the enormous monsters that swallowed Jonah and Pinocchio. At best, whales were considered benign leviathans, hidden under the sea, out of sight and out of mind. All that changed very quickly and we have, in part, the US military to thank for it.

Frank Watlington was a US Navy engineer stationed at a top-secret military base in Bermuda known as the Sofar Station. At the height of the Cold War in the 1950s, he was tasked with listening for Russian submarines approaching the coast of Bermuda, using a network of hidden underwater microphones. What he didn't expect to hear was the eerie howls and clicks of whales passing by in the distance. It was already known that whales made sounds. When ships relied on wind power, and their hulls weren't filled with the hum of an engine, whalers would often hear the faint echoes of whales from inside their wooden-hulled boats. It wasn't until Watlington was introduced to Dr Frank Payne and his collaborator Scott McVay that the true nature of these noises was discovered.

Payne, who had previously studied other animal vocalisations, quickly realised that these weren't just random grunts. The noises were structured into phrases, and the same sequence of phrases would be repeated, not just by one whale but by multiple whales. In short, they were singing. Payne and McVay were even able to identify that the songs were coming from a particular species, the humpback whale.

Payne and McVay published the first-ever descriptions of whale song in the prestigious journal *Science*. But what they did next was far more important – they released an album. In 1970 *Songs of the Humpback Whale*, comprised entirely of unedited whale song recordings, quickly sold over 100 000 copies and is still the highest-selling natural history recording ever. It contains such timeless classics as 'Solo Whale', 'Slowed Down Solo Whale' and, my personal favourite, 'Distant Whale'. Excerpts from the album even made it onto the *Voyager Golden Record*, the collection of audio recordings sent into space in the hope that one day it will be encountered by intelligent extraterrestrials.

Suddenly, people had an incredible new story to tell about whales, and people began to see them in a new light. They were no longer monsters or emotionless beasts. They were singers. People's imaginations ran wild guessing what stories were being shared back and forth between these misunderstood gentle giants. Soon the 'Save the Whales' movement would sweep the globe. By 1971 the US had outlawed commercial whaling. Australia followed suit in 1979. By 1984 commercial whaling was outlawed in all but a small number of countries. The International Whaling Commission transformed from an industrial body into a conservation authority. Whales are now protected and their numbers continue to grow. They are icons of animal conservation and environmentalism, and the discovery of whale song is widely credited with launching the modern conservation movement as we know it.

So, it *is* possible to change the way that society treats animals by learning more about them. It sounds like a big ask, to go looking for a spider discovery that will achieve for spiders what song achieved for whales, but there is no shortage of mind-blowing stories about spiders just waiting to be told. Recently some new spider species have been discovered that show promising potential for changing human–spider relations. They are very small, very talented, and thanks to digital photography and social media their popularity is growing. Unlike whales, these spiders don't sing; they dance.

Chapter 2
JUMPING SPIDERS: THE GATEWAY TO A SPIDER ADDICTION

When peacock spiders went viral

Stuart Harris had his whole life changed by a single spider. I met Stuart a few years ago when he was passing through my home town. We met up for a chat and I got to hear his story firsthand. He jokingly described himself as a 'high-acting bogan'; a lofty title he earned from years of odd jobs, from garbage collector to vineyard worker, and even a few years in the military. When not working, he enjoyed the finer things in life like red wine, fine art photography, and the great outdoors. As an avid bushwalker, he spent most weekends exploring forest trails near his home in Canberra. One weekend in 2008, when exploring Namadgi National Park, he spotted a small colourful spider sitting on a leaf. Stuart snapped a quick photo and later put it up on his personal Flickr page. Without realising it Stuart had just discovered a new species of spider, completely unknown to science.

This photo sat on Stuart's website relatively unnoticed for a few weeks until he was contacted by a local arachnologist called Jürgen Otto. A friend had forwarded Stuart's picture to Jürgen, who knew immediately that this was an undescribed species of peacock spider. Excited by the discovery, Jürgen asked Stuart if he could find another specimen. Stuart happily agreed, unaware that he was embarking on a much longer search than he had bargained for. It took three years for Stuart to find another specimen. He spent every weekend up until that point scouring the Australian bush for a tiny red and blue spider, smaller than his little fingernail. When he finally did find one, it was confirmed to be a new species and was given the name *Maratus harrisi* in honour of its indefatigable discoverer.

As their name suggests, male peacock spiders use bright, colourful displays to attract females. Unlike some other spiders who have a fat, round abdomen, male peacock spiders have a thin, flattened abdomen that is festooned with gaudy colours. When he catches sight of a female, the male peacock spider fans out his abdomen, lifts it upright and starts an elaborate dance, showcasing its dazzling colours. Male spiders shake their abdomens, wave their legs, jump from side to side and tap the ground with their feet. For millennia, these elaborate booty-shaking, toe-tapping dances have earned peacock spiders the odd tumble in the hay and, more recently, internet fame.

Combining their photography skills, Stuart and Jürgen began capturing the beauty of their newfound species. They began sharing online pictures and videos of *M. harrisi* dancing and displaying its vivid colours, and they quickly went viral. It wasn't just the spider that went viral, so did the story of the garbage-collector-turned-natural-historian who discovered it. Stuart and Jürgen became the stars of the award-winning documentary *Maratus*, which tells the story of how a chance encounter with a spider sent Stuart off on a life-changing journey. Stuart continued working with Jürgen and went on to discover more new species of jumping spiders. He is now a recognised expert in Australian spiders and has embarked on a new career in environmental science. Stuart and his namesake *M. harrisi* have their own cherished page in arachnological history, and their story has inspired many others to believe that anything in life is possible.

Since then, peacock spider popularity has continued skyrocketing. Dancing peacock spiders are now online celebrities gaining as many 'clicks' and 'likes' as any zucchini-startled cat or over-botoxed socialite. If you search for peacock spiders online, you don't have to scroll far to find videos of them dancing to flamenco guitar and the Bee Gees. Other creative filmmakers have superimposed lightsabres onto the spiders' waving legs and sombreros onto their bobbing heads. This surge in peacock spider fascination has added a much-needed boost to scientific research on these creatures. The first peacock spider was described in 1878. The published description was brief and based on a single dead specimen which unfortunately didn't give the spiders the fanfare they deserved. All this time, peacock spiders were still obscure little crawly things that popped up in museum cabinets now and then and stayed there without garnering much interest. Over the next century, the number of described species had only grown to seven.

All that changed in 2011 when the first videos of peacock spiders began appearing online. Just as song changed the fates of whales, dance has changed how much we care about peacock spiders. There are now Facebook groups and Instagram accounts dedicated solely to peacock spiders, and citizen science initiatives to catalogue and discover new species. There are now over 100 described peacock spider species. That number continues to grow as small armies of taxonomists and natural historians spend their summers crawling through the Australian bush with magnifying glasses in hand, dreaming of discovering the newest, shiniest peacock spider species. Each time a new species is discovered, it seems even journalists put aside the over-blown negative adjectives they usually reserve for stories about spiders. Instead, they write cheerily about the cute and colourful new creatures that have just been given exciting new names.

I still remember finding my first peacock spider while I was gardening in the backyard. It was a mere speck on the tip of my finger but I could make out the faint glow of a splendid iridescent colour pattern on its abdomen. I brought it inside and put it in a container while I grabbed my camera and macro lens. I took some close-up photos so I could get a better look at its glistening red, orange and blue colours, and I could finally see for myself that the famed beauty of peacock spiders was no exaggeration. Excited and utterly convinced that I had found a new species, I jumped online to see if I could identify it only to quickly find that this species had already been discovered and rather insultingly named the 'common peacock spider'.

The world knew about peacock spider dances long before the internet rediscovered them, though. In 1957, Australian naturalist RA Dunn published an article in a magazine called *Walkabout*, where he described a peacock spider he found in his backyard as a 'living jewel'. Dunn opened the article with an emphatic description: 'Surely some heathen god must have chuckled as, stealing the brilliance of ruby and sapphire, he clad that tiny form!' He went on to describe the choreography of the peacock spider's mating dance, and how sunlight glistened off the spider's abdomen, bringing to mind the 'glory of the peacock's tail'. Dunn even included a detailed drawing of the spider he affectionately named 'Hector'. It's curious that Dunn's flamboyant accounts didn't ignite peacock spider fever in the late 1950s. Perhaps because the written word simply can't do justice to the charisma of these living jewels. Peacock spiders needed modern digital photography and the volatile beast that is social media to bring their beauty to light. If you are reading this and still haven't gone online to watch videos of dancing peacock spiders, what on earth are you doing? You have my express permission to put down this book, go to the internet, and experience these miniature Baryshnikovs for yourself.

If peacock spiders are to act as a gateway for people to begin new relationships with the spider world, they won't have to look far to find some equally fascinating species. Peacock spiders belong to a group called the Salticidae, or more commonly, jumping spiders. While not all jumping spiders are as bright and colourful as peacock spiders, they are just as charming. Jumping spiders are generally small, no bigger than your fingernail, and utterly harmless to humans. Their stubby legs and compact bodies are covered in short, dense hairs; or to put it another way, they are cute and fluffy. Adding to their charm is a pair of glossy 'puppy dog eyes' that stare right at you as if to say please don't squish me. Even the most squeamish of spider haters would find it hard to resist the charms of a jumping spider.

If you encounter a spider in the wild, there's a very good chance it will be a jumping spider. There are over 4000 species of them, more than any other type of spider, and they live in every imaginable habitat, from your backyard rose garden to ice-covered mountain slopes. Many of the spider stories in this book are about jumping spiders. Given that they are the most common spider, it's astounding that people still stereotype spiders as black, sinister murder beasts. I would argue that probability dictates our typical spider should be a small, colourful, eight-legged, kitten-like, jumping spider.

The smartest spider in the world

If there were ever a spider that could compete with the fame and notoriety of peacock spiders it would have to be the legendary *Portia* jumping spiders. *Portia* spiders do not have the showy bright colours of peacock spiders – quite the opposite; at a glance they look like a small scrap of dried tree bark. But what they may lack in beauty they more than make up for with their brains. They are famous for their ingenious hunting strategies, because *Portia* spiders' favourite prey is other spiders. To hunt a fellow predator is a dangerous game. At any point, the hunter can quickly become the hunted, especially since *Portia* are small and delicate spiders, not the hulking beasts you might expect when imagining a spider that eats other spiders. Instead of using brute force, *Portia* spiders must outsmart their dangerous prey.

A few years ago, I was working in a research lab in Singapore studying the hunting behaviour of *Portia labiata*, a species common across southern Asia. We would go into the rainforest and find spiders using a rather indelicate method of collecting spiders called 'beating'. In short, you find a small leafy bush and beat it with a stick. Any unfortunate insects or spiders that happen to be in the bush get knocked off and fall into a tray held underneath the bush. As well as *Portia* spiders, we were also looking for one of their unfortunate victims, another jumping spider called *Cosmophasis umbratica*. When we found any dazed and confused spiders in our upturned umbrella, we popped them into a tiny jar and brought them back to the lab for study.

Despite both being jumping spiders, these two species couldn't be much more different. Whereas *Portia* are shaggy looking brown critters, *Cosmophasis* are shiny and colourful. Their sleek bodies shimmer with glossy black and iridescent green. Male *Cosmophasis* also reflect bright ultraviolet (UV) light, a spectrum of colours that human eyes can't see. The UV colours of male *Cosmophasis* signal their quality and are essential for attracting females, but may also be their greatest weakness as it makes them more conspicuous to predators like *Portia*. In our study we set up a lab experiment where we would give *Portia* spiders a choice between two male *Cosmophasis* spiders: one was behind a small sheet of glass, and the other behind a sheet of ultraviolet-absorbing glass that would prevent the *Portia* spider from seeing its bright UV colour patches. More often than not the *Portia* spiders would attempt to attack the spiders behind the regular glass, where the bright UV colours acted like a beacon, directing the predator towards their prey. Personally, this wasn't the most enlightening part of the experiment. Instead, I was completely taken aback just by how the *Portia* spiders looked at me. When I took them out of their jars they would often turn and stare straight at me as if to give an indignant look to the weird creature that had brought them to this strange place. It felt as if they were looking me up and down. After a while they would look around the room, glancing at the walls and ceiling above them, slowly taking in everything around them. As a scientist who studies animal behaviour, I'm acutely aware of problems with anthropomorphising animals. We shouldn't assume that animals have human-like thought processes and motivations that influence their behaviour. But the almost quizzical pose of a jumping spider looking you in the eyes and cocking its head to the side feels so astoundingly familiar it tempts you to think that there is more to that glance than we realise.

If nothing else, it was clear that my presence nearby was noticeable to the *Portia* spiders and a potential distraction that would prevent them from behaving somewhat naturally. This aspect of studying animal behaviour, called the 'observer effect' is a constant concern in biological research, as the presence of humans, especially those enthusiastically wielding clipboards and binoculars, can cause animals to act differently from the way they might if people weren't around. To avoid this, scientists must conceal themselves from the animals they are observing. Nowadays this often involves setting up video cameras to observe the animals remotely. In this case, it involved me hiding behind a thick black curtain in a dark corner of the room and watching the spiders through a small slit. The observer effect is a large concern when studying big fluffy things like mammals and birds, but it's often assumed to be less important when studying things like insects and spiders. It's unlikely that this is universally true for all insects and spiders, and the askance glances I got from curious *Portia* spiders are testament to that.

For a long time, it was assumed that insects and spiders were robotic, instinctive creatures with simple brains only capable of simple tasks. This notion has been severely challenged by more recent scientific discoveries. *Portia* spiders, with brains the size of a grain of sand, are some of the cleverest invertebrates that we know of. You've probably seen on TV or in movies the clichéd scene of a clever lab rat running through a white-walled maze to find a piece of cheese at the end. Mazes are used to study intelligence in everything from humans to slime moulds, and spiders are no exception. For spiders, mazes are usually made from suspended platforms and branches that the spiders can crawl along. And no, they don't put cheese at the end of a spider maze, they use a spider treat like a dead fly, or for spiders like *Portia*, another tasty spider.

In one series of experiments, scientists put *Portia* spiders on a high platform where they could see two lower platforms, one with food on it and one without, and two suspended walkways to choose from. The spiders were challenged to select the 'correct' route towards the platform with the food on it. The walkways weren't just simple straight lines, they were complex winding and overlapping routes. In each experiment the *Portia* spiders were able to choose the correct path more often than not, showing that the spiders were looking at the walkways, seeing where each one ended up, and deciding which one to take. No matter how complex the mazes became, the spiders could still solve the puzzle. Even when the correct walkways were longer, or initially took the spiders in the wrong direction, or out of sight of the food, or made them walk past the wrong walkway to get to the correct walkway, or when water hazards were in the way to stop them taking short-cuts, the spiders were incredible problem solvers who continually chose the correct route. Think back to when you were a kid solving mazes on a piece of paper. Did you put your pen down and just start drawing a line until you got to a dead end? Or did you look through the whole maze, find the correct route, and make a plan before you put your pen to paper? This is essentially what *Portia* spiders are capable of. They are not haphazardly wandering through mazes and solving them by accident, they are scanning the environment around them, thinking ahead, and planning the correct route before moving forward.

Using these kinds of experiments, scientists are figuring out how smart *Portia* spiders really are. They have assessed their memory and ability to remember where and what objects are, even when the spiders can't see them anymore; which implies 'object permanency', a cognitive skill that human babies don't develop until somewhere around six months of age. *Portia* spiders can count small numbers, identify different prey types after a single encounter, and solve many more complex problems. *Portia* spiders are so smart because they have to be. As mentioned before, *Portia* spiders feed on other spiders. Hunting another spider is a high-stakes game where failure doesn't just mean a lost meal, it can mean losing your life. And if my decades spent watching Schwarzenegger movies have taught me anything, it's that to hunt a hunter, you have to outsmart them. What's more impressive is that *Portia* don't just hunt one particular type of spider, they can adjust their hunting strategies to take down a myriad of different prey.

To show you what *Portia* problem-solving looks like in the wild is a bit complicated. *Portia* spiders live in complex three-dimensional environments with potential hazards and opportunities around every corner. Perhaps it's best to take ourselves into their world and go on a little adventure. I'm going to ask you to imagine that you are a spider. I would ask you to close your eyes and imagine, but that would make reading rather difficult. You are a small *Portia* spider. You are beautiful in an 'only-a-mother-could-love' kind of way, and you are about to go on an adventure. A choose-your-own-adventure. Ready your wits mighty warrior, your trials await.

THE TRIALS OF *PORTIA* – A CHOOSE-YOUR-OWN-ADVENTURE STORY

You are standing on a leaf. The morning is humid and shafts of light hang in the air as sunlight peers through gaps in the rainforest canopy. In front of you, you can see your forelimbs stretching forward. They are slender and mottled in shades of brown and white, with thick black hairs. You are a female *Portia fimbriata* in the tropical forests of northern Australia. And you are hungry.

You look up. Above you stretches a complex network of intertwining branches, vines and leaves. It is a maze you will need to navigate, and predators could be waiting behind any leaf or twig, coming from every direction. You take a few quick steps to the edge of the leaf and peer over the side. Below you, on the ground, is a thick layer of leaf litter. It provides a refuge where you could hide under leaves and rocks, but it is dark and the darkness holds many secrets. You make your first choice.

If you head up into the canopy, go to 1.

If you head down into the leaf litter, go to 2.

1.

You turn towards the stem of the plant you are on and crawl upwards. Soon you reach a fork in the path and must decide whether to go right or left. You turn your body side to side and inspect what lies ahead in each direction. The branch on the right continues for a short distance before coming to a dead end. The branch on the left continues for some distance before it turns behind a large leaf. You cannot see where it goes after that.

You go right, go to 5.

You go left, go to 6.

2.

You leave the leaf on which you are standing and crawl down the stem of the plant. Along the way, you weave between enormous water droplets that cling to the side of the plant and almost dwarf you in size. As you approach the ground, the light dims and the air turns cool. With a small jump, you are now standing on wet soil. You continue onwards, moving in and out of the darkness as you climb over and under the enormous dried leaves that blanket the earth. Now and again, you pass small worms and insects that could make for a tasty meal, but you are on the hunt for larger prey. Eventually, you stumble across something in your path: a taut string of silk. Using your forelegs, you delicately touch and smell the silk. It's from another type of spider; you know this scent and have encountered it before. The hunt is on and you begin following the thread of silk to see where it leads. You move slowly, keeping a careful eye on your surroundings until, eventually, you find your prey. In front of you stands a large black spider, *Badumna insignis*. It's not very far away and is facing towards you, but it has poor eyesight and still hasn't seen you. It is slightly bigger and much stronger than you but, if you are successful, it will make a good meal. You decide to attack.

You attack head-on, go to 3.

You approach from another direction, go to 4.

3.

You creep forward. Your powerful eyes give you an advantage and the other spider remains motionless, unable to see you, camouflaged as you are among the soil and debris on the ground. As you get closer, you ready yourself to pounce until, suddenly, the other spider moves. It takes a quick step forward and looks right at you. At this distance, it can now see you and raises its fangs ready to attack. You try to run but it's too late, the other spider leaps forward and pins you to the ground. Its strong legs hold you in place and it leans into you with all of its weight. Sharp fangs pierce your exoskeleton and venom begins pumping through your body. You die. Your adventure ends here.

4.

You turn to your right and begin walking away from the other spider. As you pass behind a large leaf you lose sight of your prey but can remember where it is and can easily find it again once you emerge on the other side. You continue until you are now directly behind the spider. From this angle, the other spider can't see you and you creep towards it. As you get close you make sure to avoid going near its legs. If you touch one of its delicate sensory hairs, your prey will be alerted to your presence. You are now right behind it, and you ready yourself for the attack. Ever so slowly you raise your fangs and extend your forelegs upwards. In one swift motion, you jump, soaring over the spider's abdomen and landing on its head. Immediately you drive your fangs into the back of your prey's head. It takes a few shocked steps forward but you hold tight and your venom works fast. Within seconds the spider slows to a halt and dies. You are victorious and feed on the nutritious innards of your prey.

5.

You take the branch leading to the right and follow it until the very end. From this vantage point, you turn from side to side and search for somewhere else to go. You can see other branches nearby and assess whether you can jump to them. Eventually, you decide that they are too far away and you will not be able to make the jump.

You have reached a dead-end, go back to 1.

6.

You take the path to the left. As you reach the bend and turn past the large leaf, you can see the rest of the path. It ends in a cluster of leaves. To your right, you see another branch not far away. You make the small jump across and land on the other branch. You continue to follow this branch until it also ends in a cluster of leaves, but attached to this cluster of leaves is a promising sign. Extending from the tip of a leaf is a taut silk thread. When you reach the end of the branch you can see the rest of the web stretching out before you. It is a large orb-web stretched between plants, and in its centre sits its owner, a large orb-web spider. It will make a good meal but you haven't encountered this type of prey before and you will need to figure out a way to catch it.

You walk away from the web and look for another approach, go to 7.

You approach the web, go to 8.

7.

You walk away from the web and explore the leaf. When looking around, you spot some more branches above you heading in different directions. One by one you inspect each branch and see where it leads. Most of them lead you away from your prey, but one of them appears to stretch over and above the orb-web. This is the route that will take you closer to your prey and you jump up towards it. As you climb along the branch, you keep an eye on the orb-web spider in the web below. You walk until you reach a point on the branch that is directly above the orb-web spider. Here is where you will make your next move. You press your abdomen against the branch and adhere a silk thread to its surface. Now that you are securely anchored you let go of the branch and lower yourself down towards your prey. Small gusts of wind cause you to swing back and forth but you continue on and slowly release silk from your spinnerets, bringing you ever closer to your prey. Finally, when you are almost touching the web below, you make your move. You leap down towards your prey and land right on top of its head. It scrambles to get away but you hold on tight and plunge your fangs into the spider, right behind its eyes. The venom takes effect quickly and the spider goes limp, hanging motionless from its web. You are victorious and may drink from the soup that was once your victim's lifeblood.

8.

Tenaciously, you take a few gentle steps onto the web. You feel it vibrate ever so slightly with each step. If you continue walking, the owner will no doubt hear your soft footsteps and you will soon become its lunch. Instead, you try something different; something bold. Using your pedipalps, the small pair of legs under your head, you strike the web. You feel the sound reverberate through the silk threads beneath you and you keep a watchful eye on the spider in front of you. It doesn't seem to react. You strike the web again, and again, but the spider still doesn't react. It can no doubt hear the sounds you are making, but perhaps they don't sound like anything it recognises. Seeing that these sounds aren't getting a response, you try something different. With your forelegs, you start plucking the silk. Immediately the orb-web spider turns on the spot and faces in your direction. Its eyesight is poor and it can't see you, but it seems interested in the sounds you are making. You continue plucking the silk and carefully watching as it takes a few curious steps forward. As you keep cautiously plucking, your prey moves closer and closer, until it stops. You don't know why. Perhaps it is suspicious. Whatever the case, the plucking signals you are making don't seem to work anymore. Now you try fluttering the silk, gently dragging the tip of one leg back and forth across the threads. Again, the spider doesn't respond. Your prey is almost within striking range but you need to try a new strategy.

You decide to leave the web and find another approach, go back to 7.

You decide to stay on the web, go to 9.

9.

Having decided to stay on the spider's web, you are in a sticky situation. You do not appreciate that pun, because you are a spider. The stakes are high and a wrong move here could mean death. Even a minor vibration could alert the spider to your presence, and the longer you stay in the web the longer you are in danger.

You continue to wait patiently, go to 10.

You take a risk and creep forwards, go to 11.

10.

You stay motionless and wait on the web. After some time, you feel a gentle breeze. Not only do you feel it with the sensory hairs that cover your body, but you also hear the web hum loudly as the wind vibrates the silk beneath your feet. Ah-ha! Keeping a careful eye on your prey, you take a few quick steps forward. It doesn't move. The noise of the wind in the web muffles the sound of your feet. When the wind stops, you stop, and the web goes silent again. When the wind picks up again, you take a few more stealthy steps. This time you take a few steps to the side and approach your prey from behind. With the sound of your approach concealed by the wind, your prey has no idea that you are now within striking range.

You leap forward and attack the leg, go to 12.

You leap over the abdomen and attack the head, go to 13.

11.

You have made the bold decision to walk, quite literally, into a trap. Each step you take, whether it is towards or away from your prey, rings ever so slightly into the taut silk at your feet. Step by step you get closer and closer to your prey until suddenly the tables turn. The owner of the web turns on the spot and runs straight towards you. You turn to flee but you are on enemy territory now, and the orb-web spider, with its long thin legs, easily outpaces you. Your frantic steps shake the web and direct the spider right towards you. The spider pounces and, with a flurry of fangs and legs, you are wrapped in thick layers of silk from which you can't escape. Your dead body is left hanging in the web, a convenient silk-wrapped snack for when the orb-web spider feels hungry again. Your adventure ends here.

12.

You leap onto the spider's leg. It was the closest part of your prey and made for a quick and unexpected attack. Your prey begins to flail madly and you have to hold on tight as you thrust your fangs into its leg. Despite the venom slowly moving through its tissues it is still alive and manages to fight back. It pivots and brings its leg, with you still attached to it, straight towards its head. It grasps with its chelicerae and pins you down with its fangs, injecting its own venomous cocktail. Eventually, it slows as your venom cripples its central nervous system, but it is too late. You and your prey both slowly die, tangled in silk and each other's legs. Your adventure ends here.

13.

You leap over the spider's abdomen and land squarely behind its eyes. Immediately you piece its exoskeleton with your fangs, injecting venom directly towards the nerve centres of your prey. Within seconds, your prey goes limp. You, mighty warrior, stand victorious on the corpse of your prey.

The truth behind the trials of *Portia*

Well, that was fun, wasn't it? Unlike some other swashbuckling choose-your-own-adventure stories you may have read, this one was not pure fantasy. Poetic licence aside, these are all actual spider behaviours that scientists have observed and studied. *Portia* spiders are incredibly flexible predators and use a range of different strategies to hunt, depending on what type of prey they encounter. Just like the spider-hunting assassin bugs we met at the beginning of this book, *Portia* will use other spiders' webs against them and pluck silk as a bluff to draw out their prey. Rather than relying on a specific type of vibration, they will try different types of vibrations and use trial-and-error to pick one that seems to work. *Portia* will use monotonous vibrations that seem to desensitise their prey to the sounds being made in their web, or auditory 'smokescreens' to sneak up on their prey, taking quick steps forward while the sound of their footfalls is muffled by wind.

Since *Portia* live in such complex three-dimensional environments and are faced with demanding problems with potentially lethal consequences, is it any wonder that they can easily solve mazes in a research laboratory? This knack for maze-solving comes in handy when hunting, and *Portia* will explore different ways of approaching their prey that will either lead to better capture success or less chance of them becoming prey themselves. This is also dependent on the prey. For example, when hunting something like a daddy-long-legs spider they will sneak through gaps in between their long spindly legs, while something like a house spider requires *Portia* to sneak around and attack from behind. More often than not, *Portia* will direct their attack towards the head, meaning that the venom acts much more quickly as it immediately comes into contact with the victim's brain. Studies have even challenged *Portia* by presenting them with spiders that they have never encountered before, or that might be from a completely different country, and *Portia* can masterfully figure out, on the spot with no preparation, how to defeat them.

These hunting behaviours are less like those of other spiders and are similar to what we see in large predatory mammals like cats and dogs. These behaviours require types of intelligence not usually associated with small invertebrates, like learning from trial-and-error, or thinking ahead and coming up with a plan before acting on it. If *Portia* can plan ahead, does this mean they can envisage future scenarios that inform their decisions? And does this imply that they have imaginations? Can spiders dream? I know what you're thinking. *Really James? Spider dreams? Get real buddy.* Well, wait till you hear this. Jumping spiders have eyes that can slightly move from side to side and in 2022 scientists discovered that the jumping spider *Evarcha arcuata* enters a sleep-like state while resting at night. They hang upside down from a strand of silk and, while asleep, they twitch and curl their legs, and make 'rapid eye movements', or REMs. Much like humans and other vertebrates, these spiders seem to go into a REM sleep state. This sleep state is believed to be linked to dreaming in humans, where the rapid eye movements are made in response to whatever the person is 'seeing' in their dreams. Maybe spider dreams aren't such a far-fetched idea after all.

We still don't quite understand how *Portia* manages to coordinate complex behaviours with such a small brain. In the past, animal intelligence has been linked to overall brain size – the bigger the brain, the more brain cells you have, and the more neuronal connections that can be used to do complex calculations. Smaller brains are limited because they must have fewer brain cells. Despite this limitation, *Portia* species continue to astound scientists with what their small brains are capable of. They challenge what we thought we knew about animal intelligence, neuroscience, and computational thinking. And it's not just their brains that make jumping spiders such remarkable predators and problem solvers; their eyes are some of the best in the animal kingdom.

Eyes like a hawk (literally)

The glossy, forward-facing eyes of jumping spiders aren't just there to make them look cute; they are miniature telescopes that give them the sharpest vision of any small animal in the undergrowth. Unlike other spiders that catch prey in webs, jumping spiders rely on keen eyesight to stalk and pounce on their prey. At a glance, jumping spiders can navigate complex environments, tell the difference between males and females of their own species, and spot different types of prey. And of all the jumping spider species that have been studied, the *Portia* genus comes out ahead once again. Not only is their eyesight better than that of any other spider, it's better than that of any other small insect, and most large vertebrates. In the insect kingdom, dragonflies are renowned for their incredible eyesight. In some dragonfly species, enormous compound eyes wrap around their entire heads, and can be made of over 30 000 individual light-sensing organs per eye. Even so, the sharpness of *Portia*'s vision beats dragonflies' by an order of magnitude.

Unlike insects that have compound eyes made up of many separate light-sensing ommatidia, spiders' eyes are more similar to our own. They have a cornea and lens that focuses light onto a retina. But while our eyes are spherical, jumping spiders' eyes are long tubes that extend back into their heads. They focus light at multiple points along the tube; first at the cornea and lens, and then again at a secondary lens – the 'concave pit' – that magnifies the image onto a layer of dense retinal cells. This means that their eyes are, quite literally, telephoto lenses that are specialised for focusing on objects far off in the distance. The image that hits the jumping spider retina is as about as sharp as the laws of physics will allow. The retinal cells are tightly packed: spaced about one micron apart. If the retinal cells were packed any tighter the width of each cell would start to become smaller than the wavelength of light photons.

The eyes of jumping spiders that can focus on faraway objects are somewhat similar to the eyes of predatory birds like falcons and hawks. These birds have the sharpest vision of any animal on the planet and can spot small animals on the ground while they fly high above them. This is because they have their own 'telephoto' eyes, with a secondary lens at the back of the eye that magnifies images onto a field of densely packed retinal cells. Since jumping spiders have similar structures in their own eyes, it's not just a figure of speech to say that they have 'eyes like a hawk'.

In addition to their super-sharp central eyes, jumping spiders have three extra pairs of secondary eyes that run down the sides of their head. These eyes aren't as good at forming images as their central eyes, but they give the spiders wide panoramic vision. For some species, it wouldn't be a stretch to say they have eyes in the back of their heads.

While jumping spiders can move their retinal tubes side to side slightly, like during REM sleep, they can't really turn their eyes to look around. To scan their environment, they have to move their entire bodies, focusing on a single point at a time. This probably explains their jerky nodding and head-cocking behaviour that makes them seem so inquisitive and endearing. And if you haven't fallen in love with jumping spiders yet, then I have one more story that I hope does the trick. It's about a spider that seems to be using its keen vision for the benefit of humanity. And it's the only spider we know of that feeds on blood. Well ... kind of. It's complicated.

Spiders *vs* mosquitos

If there was ever another animal that has suffered more public disdain than spiders, it would be the mosquito. This isn't all that surprising given that they feed on human blood, and in so doing can spread diseases like malaria, dengue, Zika virus, filariasis, and many other serious illnesses. Malaria is a horrible disease that is preventable and curable, yet still widespread and responsible for hundreds of millions of deaths worldwide. It's caused by single-celled *Plasmodium* parasites that multiply rapidly inside the human bloodstream. When mosquitos bite humans and other mammals they can ingest blood containing the *Plasmodium* parasite. The parasite continues its development inside the mosquito before being passed on to another host through the salivary glands of the mosquito. Not all mosquito species are vectors for malaria. Only mosquitos in the genus *Anopheles* are known to transmit malaria-causing parasites.

An estimated 96 per cent of all malaria deaths worldwide occur in Africa. Eighty per cent of these deaths are among children under the age of five. The threat that malaria poses to Africa was no doubt in their minds when scientists discovered an African jumping spider that specialises in eating mosquitos. What makes these spiders even more special is that their preferred prey are *Anopheles* mosquitos. And, if that's not surprising enough, they target *Anopheles* mosquitos that are full of fresh red blood. *Evarcha culicivora* can be found munching away on mosquitos on the shores of the great Lake Victoria. The shores of this tropical lake are teeming with insect life, so this spider can take its pick of any prey, but specifically goes after *Anopheles* mosquitos. Even seasoned entomologists can struggle to distinguish different mosquito species, but the sharp eyesight of *E. culicivora* allows them to distinguish *Anopheles* from other species. If in doubt, the spider can also follow the scent of an *Anopheles* mosquito. When standing on the ground, *Anopheles* tend to hold their hind legs off the ground and point their abdomen high in the air. This characteristic posture also makes them vulnerable to attack from spiders. When stalking an *Anopheles* mosquito, *E. culicivora* will use a special strategy where they sneak up from behind and attack the exposed upheld abdomen. If the mosquito takes off, the spiders have to hold on tight, riding their prey through the air as they begin feeding on the mosquito and its bloody insides. This attack strategy is most common in juvenile spiders, which can take down mosquitos many times their own size.

Evarcha culicivora prefers female mosquitos over males, probably because male mosquitos don't feed on blood and won't come with the added benefit of a tasty blood-meal. Once again, the spiders' sharp eyesight comes in handy for telling the difference between the small antennae of female mosquitos and the thick bushy antennae of males. The spiders can also pick up on the red hue of a blood-filled mosquito abdomen.

It's possible that *E. culicivora* gets additional nutrients when feeding on blood-filled mosquitos, but there is another strange reason why these spiders have a thirst for blood. It helps them get laid. There is nothing more alluring to these jumping spiders than the smell of a jumping spider of the opposite sex that has just eaten a big, bloody mosquito. It's not just that they like the smell of the blood; there is something particular about the bouquet of spider pheromones that comes from a blood-fed spider of the opposite sex. The effect is only temporary, though. In lab experiments, scientists showed that the attractiveness of spiders declined the longer they went without feeding. Perhaps the scent that comes with having fed on a blood-filled mosquito indicates their nutritional condition or their status as a top-notch hunter, and that makes them a more desirable mate. So, not only must *E. culicivora* obtain this alluring perfume by feeding on mosquitos, they have to continually replenish the scent by feeding on even more mosquitos.

Could you ever imagine a better ally for humanity than an adorable jumping spider whose life purpose is to eat as many mosquitos as possible? Luckily for us, *E. culicivora* seem to like people too. They are often found in and around houses, and they even seem to like our smell. In a clever little experiment, scientists presented jumping spiders with the scent of either a nice clean sock or a sweaty sock that had been worn for a day. They discovered that, unlike your housemates, these jumping spiders love the smell of your sweaty socks. *Anopheles* mosquitos also use the smell of sweat to find their way toward humans, so jumping spiders are likely homing in on this same cue to find their prey; because where there are sweaty humans there will probably be some tasty blood-filled mosquitos.

Since the discovery of *E. culicivora*, another mosquito-hunting jumping spider has been discovered hiding inside fallen bamboo stalks in Malaysia. As bamboo decomposes on the forest floor, the hollow chambers that comprise the stalk inevitably start to fill with moisture and rainwater. These small stagnant water bodies are the perfect habitat for aquatic mosquito larvae. Which also makes them the perfect habitat for *Paracyrba wanlessi*, a jumping spider whose favourite food is mosquitos. Unlike *E. culicivora*, who generally only hunt adult mosquitos, *P. wanlessi* preys on mosquitos at all life stages (adult, pupae and larvae) and prefers them over any other type of prey.

The discovery of these two mosquito-hunting jumping spiders has generated a bit of excitement as to whether they could be used as biocontrol agents to help stop the spread of mosquito-borne diseases. As exciting as this possibility is, evidence is currently lacking. The World Health Organization has yet to turn its resources towards mass rearing malaria-busting spider armies and instead continues to focus on the tried and tested approach of providing antimalarial drugs, netting and insect repellent. At the least, we can say that these spiders pose absolutely no threat to humans, while potentially benefiting us by hanging around our homes. Later in this book, we'll discover that there are other benefits to having the odd spider or two running around the house. For now, I hope these jumping spider stories have started you on your journey to seeing spiders in a new light.

Chapter 3

SPIDERS IN DISGUISE

Spiders aren't the only cunning predators traipsing through the leaf litter. In this micro-world there are formidable predators everywhere: centipedes, praying mantises, tiger beetles, assassin bugs, velvet worms, the list goes on. Not to mention the parasitic wasps trying to lay eggs inside you and the mind-controlling fungi that explode out of your skin. Life is tough for the little guys scurrying through the undergrowth. But of all these miniature foes there is one that strikes fear into the hearts of almost all small animals – ants.

A single ant is not so threatening but, for small creatures, messing with an ant is the insect equivalent of picking a fight with a Hells Angel. Because where there is one ant, there are bound to be more, and they don't take kindly to folk messing with one of their sisters. In large numbers, ants are a deadly force that can swarm and take down animals that otherwise dwarf them in size. Ants are strong enough to lift many times their own body weight, and can be armed with powerful jaws, painful stings and formic-acid sprays. For their own safety, most spiders and other predators will avoid trying to eat ants in the first place.

If you have gone bushwalking in the tropics and have and accidentally stumbled across a nest of green tree ants, you probably have vivid and painful memories of it. Green tree ants (*Oecophylla smaragdina*) are also known as weaver ants because they construct their nests out of leaves woven together with dense strands of silk. The nests are suspended from an overhanging branch and, in my experience, usually hang menacingly about head height. When hiking through the rainforests of far north Queensland I discovered I had an aptitude for finding green tree ants' nests, using a simple method of not watching where I was going. If I paid too much attention to where I was putting my feet, soon enough my head would collide with a large ants' nest, triggering an immediate attack. Within a matter of seconds, my entire torso would be swarming with tiny yellow-green ants. Thousands of needle-like stings and pinches would spread across my skin as the swarm began defending themselves against this clumsy bushwalker. The only sensible response is to begin wildly flailing your limbs and gyrating about in a kind of frantic go-go dance. Eventually you resort to frantically stripping off pieces of clothing to rid your body of the crawling swarm of miniature assailants.

Every now and again I was lucky enough to spot an ants' nest before colliding with it. If you manage to do this, take the opportunity to spend some time observing the ants. You can get quite close to them and watch as they crawl in and out of the basket-like surface of their nest. When they're not attacking you, green tree ants are quite beautiful creatures. And if you are lucky, you might spot an ant that looks a little bit different to the others. It's different because it's not an ant at all, but a spider.

Ant-mimicking spiders

The Australian jumping spider *Myrmarachne smaragdina* is a master of disguise that has managed to join the ranks of one of the toughest ant-gangs on the block. Their shiny exoskeleton and yellow-green bodies make them almost indistinguishable from a green tree ant. Their heads and abdomens are 'pinched', creating a waist that matches the telltale segmented shape of an ant's body. Two large black dots on their head make for convincing 'false eyes'. As they move about, they run in irregular bursts with exaggerated leg movements and wave their front legs out before them, effecting a near-perfect impression of a wandering ant waving its antennae.

Now, imagine that you're a predator on the hunt for some tasty little spiders. You come across this small yellow creature that looks just like a green tree ant. Knowing how aggressive and persistent these ants can be, would you be willing to take a chance and mess with it? *Myrmarachne smaragdina* is just one of over 300 species of spiders that look like ants. By masquerading as miniature Hells Angels, these relatively harmless spiders can trick other larger predators and avoid being eaten.

Spiders aren't the only critters that have jumped on the ant-mimicry bandwagon. Ants have such a hard-core reputation that there are praying mantises, wasps, bugs, crickets, beetles, even plants that have all evolved strange ant-like disguises as a means of protection. So next time you find yourself thinking that spiders are scary creatures, remember that, as far as nature is concerned, spiders aren't that bad. It's the leather-clad, motorcycle-riding, formic-acid-spraying ants that are hell on six legs.

For some spiders, the predators they are trying to avoid are other spiders. Even keen-eyed, clever jumping spiders like *Portia* struggle to distinguish between ants and ant-mimicking spiders and will avoid attacking the ant-mimics in favour of regular-looking spider prey. One species of ant-mimicking spider takes things to a whole new level. *Mymarachne melanotarsa* use their ant-like appearance to feed on other spiders. Their disguise strikes such fear into other jumping spiders that spider mothers who have recently laid eggs will completely abandon their nests at the sight of an ant-like *M. melanotarsa*, leaving their offspring easy prey for the ant-mimic.

If ants are the toughest critters on the block, spiders should want to avoid them too. Which raises the question as to whether ant-mimicking spiders are good enough to fool ants. The short answer is yes, of course. Spiders are awesome. Even with aggressive swarming ants, like green tree ants, there are several different species of spiders that can be found quite happily wandering among the ant colony. Ants don't have the best eyesight, so even an imperfect ant-mimic might be good enough to make for a passable ant. But what ants lack in eyesight, they make up for with their sense of smell. Ants live in a chemical world: they use scents to mark their trails and guide their nestmates to and from food sources. Every ant is covered in a scent unique to their individual colony, which they use to distinguish nestmates from intruders. When one ant detects an intruder, it can release an airborne alarm pheromone calling nearby ants to defend the nest. These ants also release their own alarm pheromones, drawing in even more attacking soldiers. A spider looking to fool an ant doesn't need to just look like one, it needs to smell like one.

For this reason, there can be ant-mimicking spiders that,

to our eyes, don't even look like ants. The jumping spider *Cosmophasis bitaeniata* doesn't look like an ant at all, it looks like, well, a spider. And yet, it survives inside green tree ant nests. Australian scientists analysed the chemicals on the surface of these spiders and found that they matched those of the ants in the colony, so that if an ant detects the scent of the spider, it smells like just another nestmate. It's safe to assume that by seeking refuge among the colony and within the nest, the spiders are well protected from any predators, including the ants themselves.

Be warned, this story doesn't end here. If you would prefer to sleep soundly thinking about spiders and ants living peacefully side by side then I suggest you skip the next two paragraphs. There's more to this relationship and there is more in it for the spiders than just protection. Despite the hospitality of the ants, this spider breaks the number one rule of being a good house guest: don't eat the babies. *Cosmophasis bitaeniata* feeds off the larvae of green tree ants, and by using its chemical camouflage it can walk right through the house into the nursery and take babies straight out of the arms of the nursery maids. Scientists have discovered that the spiders can simply walk up to one of the worker ants that is carrying an ant larva in its mandibles and just take it from them. The ants don't even put up a fight, they seem to just hand the larva over to what they believe to be a nestmate. The spider then happily walks off munching away on the little ant baby.

Ant larvae are the favourite food of the spider, and possibly a crucial part of their diet. When scientists tried raising *C. bitaeniata* spiderlings in the lab, the spiders didn't survive unless they were fed at least some ant larvae. Not only is it crucial for their general survival, but it is crucial for maintaining the spider's ant-like disguise. The spiders can't make ant-mimicking chemicals themselves; they get them from eating ant larvae. By feeding on baby ants these spiders take on the scent of that particular colony, either by sequestering the chemicals into their tissues, or maybe just by being messy eaters and wiping baby guts all over themselves. Presumably, the more larvae the spiders eat, the better ant-mimics they become.

Spiders are such convincing ant-mimics that one particular species appears to have gotten cocky; it's not content with just looking like an ant, it shows off by doing it backwards. Unlike other ant-mimicking spiders, whose heads tend to resemble ant heads and whose abdomens resemble ants' abdomens, the Malaysian *Orsima ichneumon* does it in reverse. The top of the spider's head is glistening green, like the iridescent abdomen of an ant. And the spider's rear end is long, with a bulbous tip, giving the impression of a rounded ant's head. It's spinnerets, which are small and inconspicuous in other spiders, are long and dark black. In an old natural history paper from 1976, Professor Jonathan Reiskind first put forward the idea that these spiders could be mimicking a backwards ant and describes how one long pair of spinnerets waved about like antennae, while another two shorter pairs of spinnerets moved side to side just like a pinching set of ant mouthparts. Admittedly, *O. ichneumon*'s backwards illusion takes a bit of squinting to make out, but in nature, where animal sensory systems are so different to our own, this strange disguise might just be all that this jumping spider needs to convincingly pass for an ant.

Some good friends of mine actually made an unexpected discovery about how *O. ichneumon* use their deceptive skills when their original research plans hit a literal roadblock. A team of scientists led by Dr Christina Painting was driving through the rainforests outside of Kuala Lumpur on the way to a study site that they knew would be a good place to collect spiders. On the way they discovered that an overnight landslide had blocked the only road through the forest. Having already been driving for a while, the team decided to take a break and explore the roadside vegetation. They found a few small shrubs that had leaves covered in ants busily feeding on nectaries – small wells in the leaf surface where the plant secretes sugary nectar. These nectaries keep the ants around, and they then provide the plant with some muscle power that can keep caterpillars and aphids at bay. Being keen spider hunters, the scientists quickly spotted the odd backwards-walking ants that were, actually, forwards-walking spiders. As if to add to their ant-like disguise, the *O. ichneumon* spiders were also walking between the nectaries and feeding on the sugary liquid. This wasn't all that surprising, given that we know of other spiders that will opportunistically feed from nectaries, but what happened next had never been seen before. After feeding on the nectar the spider turned around and, using its abdomen, wove a silken shield over the top of the nectary. It then continued on; feeding from other nectaries before covering them each in silk. It looked like each spider was patrolling a leaf, protecting its nectaries from wandering ants. When smaller ants approached, the spiders would chase them away. If larger ants approached the spiders would have to hide, hoping that their silken shields were enough to stop the ant messing with the food. I can't help but admire the sheer audacity of this little jumping spider, boldly eking out a living on ant-infested plants.

That this discovery was only possible because of a landslide that disrupted the scientists' original plans is an example of how unpredictable scientific advancement can be. Scientists are employed to test particular hypotheses or answer specific questions. Obviously, this means focusing on a set problem to reach a specific answer, but what can be forgotten in this pursuit is that there is still a wide world out there, full of answers for which we haven't even discovered the questions. Taking time out to just observe and explore, without a set goal or the expectation of measurable outcomes, is sometimes the best way for new and unexpected discoveries to find their way to you. With the pressure that is put on scientists to always deliver results and data in the shortest and most cost-effective way possible, it sometimes requires an earthmoving force, like a landslide, to give them an opportunity to just sit back and watch the world go by.

To be honest, I am envious of the ingenuity and calm collectedness shown by my colleagues, because I can testify to the peculiar dangers of this specific stretch of road in Malaysia. A few years before this expedition, I spent some time doing research in the same forest and would take the same route up and down the mountain. It's a narrow, winding road that snakes its way through the wilderness, eventually coming out at the Genting Highlands – a dense cloud forest where you'll find huge tropical butterflies, endangered dusky langurs, and rare and delicate pitcher plants; not to mention a 7000-room hotel and casino complex complete with indoor theme park and 18-hole golf course.

My time in the highlands was spent avoiding the casino and exploring the forest instead. Like my colleagues, I would also spend quite a bit of time stranded on the side of the road, although none of my misadventures would turn out to be as productive as theirs. One time my hire car broke down halfway up the mountain and I spent the next five hours waiting for a tow truck to arrive. Another time I was stranded on the roadside was when I and two colleagues were in a car following a line of trucks up the mountain. A car that was coming down the hill hit the muddy shoulder of the road, spun out of control and collided with our car. Luckily, no-one was seriously injured but both cars were un-driveable. Again, I was left sitting at the side of the road waiting for a tow truck to arrive, but at least this time I had company with me. Having survived a pretty serious car crash, we decided to sit quietly and contemplate our own mortality rather than go hunting for spiders in the undergrowth. I'm not sure what the lesson is here. Maybe it's something to do with taking time out to observe the natural world, or thinking quickly and making lemons into lemonade. Or maybe the lesson is to just skip the scenic route altogether and take the main road to the casino for a cocktail.

One thing I did manage to pay attention to while stranded in the rainforest was the abundance of *Asystasia* flowers that littered the roadsides. I spent countless hours watching the constant flow of bees and flies coming to visit these small white and purple flowers. Yes, because I was going stir crazy waiting for tow trucks to arrive, but also because it was part of the research project I was doing. Every now and again I would find, hidden inside the tiny white flowers, an even tinier crab spider. They were an impossibly bright white and would almost disappear among the white petals of the flower.

Crab spiders and other ambush hunters

Crab spiders demonstrate beautifully how camouflage can be used for hunting as well as protection. These spiders can vary in colour from bright white to rich yellow, sometimes with hints of pink or light green. While their vision isn't great, they are able to see colours and choose a flower to sit on based on its hue. By hiding in flowers that match their colour, all these spiders need to do is wait patiently for a pollinator to arrive and then pounce. The bright flower colours do all the work for them by attracting prey right towards the spider.

In 1882 one James Angus wrote a short letter to the journal *The American Naturalist* with a one-paragraph description of 'the little flower spiders' that he had observed. Over a century ago, Angus made a surprising observation, saying, 'If I take a white one and put it on a sun-flower, it will get quite yellow in from two to three days.' This is one of the earliest mentions of crab spiders in the scientific literature and it established the incredible fact that crab spiders can change colour to match their surroundings. If the flower dies or bees become savvy to the presence of a predator nearby, the spiders can move on to a new flower and change colour to match their new surroundings. The spiders do not shed their skin to change colour: we now know that a spider's hard exoskeleton is made of living cells that contain a variety of pigments. By either producing or breaking down different pigment types, the spiders can change colour in a relatively short time. This raises even more questions about how spiders interact with flowers. How does the surface the spider sits on trigger the metabolic reactions that change their colour? Does seeing the flower colour dictate what colour the spider changes to? Does the spider know what its own colour is? Is that even a relevant question, because we don't know if spiders have a concept of *self* in the first place? These head-scratchers start reaching into the philosophical and are still waiting to be answered.

There are endless ways that spiders can disguise themselves or blend into their backgrounds. Bird-dropping spiders masquerade as (you guessed it) a glistening white-and-black droplet of bird poo. *Poltys* spiders are near impossible to find when they are hidden along a branch; their long jagged abdomens stick up in the air and look just like the broken stump of a twig. Wrap-around spiders (*Dolophones*) have flattened bodies that literally wrap around the curves of tree branches. With mottled skin the colour of bark or lichen, they virtually disappear in front of your eyes. Just like crab spider disguises, these aren't just means of protection from predators. They can help spiders catch prey as they lie camouflaged waiting to ambush unsuspecting insects. In some cases, their disguises actually lure prey towards them. Take bird-dropping spiders for example: it's easy to understand how this scatological charade protects spiders from predators who don't want to risk getting a mouthful of faeces. But it also attracts animals who actually *do* want a mouth full of faeces; by which I mean flies.

Recently, scientists in southern China studied the bird-dropping spider *Phrynarachne ceylonica*. This spider is actually a close relative of flower-dwelling crab spiders, but instead of shining bright in pretty pink, these spiders are mottled in a glossy white and black bird-poo pattern. They discovered that bird-dropping spiders actually attracted flying insects towards them, like, as the saying goes, flies on shit. When the scientists changed the colour patterns of the spiders by painting them either all-white, or all-brown, they became less attractive to flies, meaning that the spiders aren't just attractive because they are a conspicuous colour, but that they also need the right ratio of white to dark to look like a most delicious pile of poop.

Even white crab spiders, who are considered classic examples of animal camouflage, can sometimes be more than just passive predators waiting to ambush their prey. You would think that animal camouflage research would be a pretty intuitive field of science; if an animal is hard to see then it must be camouflaged. End of story, right? Not quite. Remember that humans are just another animal with senses that are unique to us. We don't have the sense of smell that ants do, and we can't listen to vibrations the way spiders do. What we do have is excellent colour vision, but even then our vision has its limits. There are colours outside of the human visible spectrum that we cannot see, but other animals can. Ultraviolet sits just below blue on the electromagnetic spectrum and it's actually a little strange that humans can't see UV when so many other animals can. The only way for us to study how animals use UV light is with the aid of technology.

A team of Australian scientists studying the crab spider *Thomisus spectabilis* used a spectrophotometer to measure what colours the spiders reflected and how other animals might perceive them. To us these crab spiders look like any other, a bright white spider on a bright white flower. But when the scientists measured the spiders' colours they found that they shone brightly in the UV part of the colour spectrum. This on its own is not all that unusual; lots of things reflect UV light. What was strange about this revelation was that the flowers the spiders were on *didn't* reflect UV light. So, from the perspective of something like a bee, these spiders aren't camouflaged at all; they stick out like sore thumbs. And yet, they are very effective hunters and catch bees despite the fact that they aren't camouflaged against the flower. The scientists conducted an experiment where they gave honeybees the choice between a white flower, and a white flower with a conspicuous UV spider on it – the bees actually preferred the flower with the predator.

To dive further into this, the scientists designed a very simple, elegant experiment. They painted the spiders with a UV-absorbing chemical, generally used as an ingredient in sunscreen, to stop the spiders reflecting UV light. When they removed the UV colours, making the spiders more camouflaged, they were less attractive to bees. It turns out that bees actually like flowers with contrasting colour patterns, which is why many flowers are often not just one colour, but show off bold patterns to get the attention of pollinators. By reflecting bright UV, these Australian crab spiders created a big bright bullseye in the middle of the flower, and actually enhanced its attractiveness.

One crab spider has managed to take flowers out of the equation altogether and lures prey without ever having to sit on a flower. The South American *Epicadus heterogaster* can sit exposed on a green leaf, nowhere near a flower, and still attract insects. Like other crab spiders, its body shines bright in hues of white, yellow and pink. Unlike other crab spiders, its abdomen has short spikes that radiate outwards. At first glance, it looks like a small flower and as far as some insects can tell, the spider *is* a flower and they will curiously fly towards it, unaware of the dangers. A team of scientists from Brazil studied *E. heterogaster* to see if UV light was an important signal, as in other crab spiders. Similarly, they delicately painted the spiders with sunblock and discovered that when UV was removed, insects stopped showing any interest in the crab spiders.

I love hearing stories about how scientists can make discoveries using simple tricks, like putting sunscreen on a crab spider, or colouring them in with water-based paint, and then watching to see what happens. Yes, we all know that science can involve lasers and rockets and supercomputers but sometimes, to answer a question, simpler is better. You can make incredible scientific discoveries with nothing but playdough, paper clips, a pen and paper, and the ingenuity to know the right questions to ask. I still remember as a university student listening to a lecture about the Australian crab spiders that changed the way we think about animal camouflage, and how a simple experiment with sunscreen beautifully demonstrated the limitations of human eyes. The person giving the lecture was one of the scientists who made this discovery, Professor Mariella Herberstein. She would soon become my research supervisor, and is still an important mentor and friend. After finishing my undergraduate degree, I joined her spider research lab, and would be sent off on adventures into the field looking for weird and wonderful critters; the first of which was a trip to far north Queensland in search of small tree-running praying mantises, where I also learnt that I had an uncanny knack for finding green tree ant nests with my head.

The spinning bolas spider

There are seemingly endless stories about the different disguises and deceptive tricks that spiders can play on other animals. This chapter would start to get pretty tedious if we attempted to go through them all, but I have saved the most famous case of spider trickery for last, because I would argue that it's also the most bad-ass. If ant-mimicking spiders are wannabe Hells Angels, then bolas spiders (genus *Mastophora*) are wannabe cowboys. They mostly feed on male moths. This presents a problem because, as you know, moths can fly and spiders don't. So, instead of going after the moths, bolas spiders have evolved a way to bring the moths to them.

Male moths fly around at night using their antennae to sniff out females. With romantic intentions, they detect airborne chemical trails and follow them to find lady moths. Now and again, however, a poor male is duped into following a different scent. At night, bolas spiders will perch on the end of a twig or a branch to catch their prey. When ready, they release an airborne chemical cocktail that matches the pheromones of female moths. Male moths tricked by this fake pheromone will fly towards the waiting predator, but this is only the first half of the spider's ploy. The next step is to catch the moth in a silken trap.

The bolas spider spins a thread of silk with a heavy glue droplet on the end. Using one of its legs, the spider will start spinning its silken 'bolas' in circles above its head. Imagine an eight-legged cowboy spinning a silk lasso in the air and you've got a pretty good image of what this looks like. The spinning circles get wider and wider until a poor love-struck moth collides with the silk and gets wrapped in sticky tendrils. The bolas spider then crawls down towards its prey dangling at the end of the weaponised web. The moth is wrapped and held in place with even more layers of silk, where it will spend its last moments waiting for death, no doubt regretting its over-eagerness and mourning the love that could have been.

Chapter 4
THE HUNTER'S ARSENAL

The vegetarian spider

In the forests of Mexico, you will find thorny acacia bushes crawling with *Pseudomyrmex* ants. These ants are not an infestation, they are living in a complex symbiotic relationship with the plant, acting as sentinels and caretakers for their living home. As they walk along the leaves and stems, they hunt and protect the plant from leaf eaters and sap suckers. They clean up pesky mould and clear away any vines that might try and twirl their way around the acacia bush. In return for these services the plant makes a welcoming home for the ants. The thick thorns of the acacia bush are actually hollow caverns that form a tiny neighbourhood where ants can take shelter and care for their larvae. And, as if the ants needed another reason to stick around, the acacia plant actually provisions out a balanced diet for the ants. Scattered along the branches of the acacia bush are little droplets of rich sugary nectar, and on the tips of its leaves the plant grows specialised structures called 'Beltian bodies' that are rich in fats, proteins and amino acids. With shelter and food in ready supply, the ants hardly ever need to stray from the plant and can be full-time security guards.

Beltian bodies are unusual structures; they look like small yellow pods budding from the green tips of the acacia bush leaves. Leaves aren't usually rich in fats and proteins, but the acacia bush concentrates its resources into these small packets specifically to keep its ant friends close by. They are also unusual in that Beltian bodies aren't just a food source for ants, they are also the favourite food of the world's only vegetarian spider.

The delightfully named *Bagheera kiplingi* expertly ducks and dodges its way around the acacia bush, avoiding the ant security guards, picking off the nutritious Beltian bodies. Every now and again it will also take a sip from the nectar droplets scattered across the plant. Being spiders, they are partial to snacking on the odd ant larvae when the opportunity presents itself, but on the whole this spider's diet is almost entirely vegetarian. Studies show that over 90 per cent of the spider's diet comes from acacia bush Beltian bodies.

In case you are wondering, yes, *B. kiplingi* is named after Rudyard Kipling and his character Bagheera from *The Jungle Book*. George and Elizabeth Peckham (a spider taxonomy power couple from the late nineteenth century) had a penchant for naming spiders after Rudyard Kipling characters. We can only guess what inspired them when naming this particular species, but the wise and powerful panther Bagheera is a fitting namesake for the cunning *Bagheera kiplingi*.

So, you could take away from this story that not all spiders are fearsome bloodthirsty carnivores; but let's not beat around the acacia bush here – spiders are predators, and amazing ones at that. With the possible exception of *Bagheera kiplingi*, all spiders must eat other animals to survive, and the incredible ways that they do it are just another part of what makes them awesome. Combining feats of stealth and weaponry, spiders are some of nature's most exciting predators.

There are impressive and verifiable reports of spiders taking down relatively large vertebrate prey. Spiders that live near water have been observed catching fish over twice their own size. There are accounts from people who have watched spiders catch fish from their backyard ponds. Spiders that build large webs can ensnare the odd bird, bat or snake, and there have been a few observations of ground hunting tarantulas catching the occasional frog or mouse. As impressive as these feats are though, they are rare, uncharacteristic, and definitely the exception rather than the rule. Even the famous 'goliath birdeater' (*Theraphosa blondi*), the world's largest spider by mass, rarely eats anything other than insects and worms.

When spiders are on the hunt, they have two main weapons in their arsenal: silk and venom. Using one or a combination of these weapons, spiders can catch all manner of prey in either very simple or weirdly elaborate ways. Often, silk is used first to capture their prey, and venom is used second to kill their prey. Let's start by exploring some of the different ways that silk is used as a trap, and then get into the science of how and why venom works.

Silk: The ultimate hunting tool

There is an entire chapter later in the book dedicated to the incredible substance that is spider silk. So, for now let's just say that spider silk is incomparable to any other natural material for its strength and flexibility. It can be custom built by the spider for different material properties, and woven into all manner of traps, snares and pits for catching prey. The complexity of the silken traps that spiders build tends to follow a pattern with spiders' evolutionary tree. The more 'modern' araneomorph spiders are more likely to use elaborate prey-capture webs, while the more 'ancient' mygalomorphs rely on silk as a sensory tool to help find prey. There are of course many exceptions to this broad pattern, but in general mygalomorphs tend to rely on stealth and brute force to subdue their prey.

Ambush predators

You'll remember that mygalomorph spiders are the big hairy ones with fangs that point downwards. These tarantulas, funnel-web and trapdoor spiders are also known to be pretty sedentary, and during the day are hidden away in burrows, under rocks or in crevices. At night they will come out to hunt, but to be fair they aren't exactly go-getters. While some will come out of hiding for a slow wander about, many of them will simply take a few steps to the entrance of their burrow where they will wait for prey to pass. These large spiders are ambush predators and will wait for a large insect, or maybe even another spider to pounce on. Many of these ambush predators never feel the need to leave their hiding places. The Brazilian tarantula *Pachistopelma rufonigrum* makes its home inside tank bromeliads – large plants with broad leaves that form a well of water in the centre. Here the tarantula can hide and wait for other animals that come to use the water inside the bromeliad, like frogs, salamanders or aquatic insects.

Other ground foraging spiders will sit at the entrance to their burrow and watch and listen for prey. Here silk comes in handy as it can augment the spiders' ability to hear their prey approaching. Perhaps the most famous of these ambush predators are the trapdoor spiders. Their burrow openings are completely hidden behind a little door made of dirt and debris. The doors are often hinged with a layer of silk on one side that will pull the door closed as the spider retreats back inside. When prey comes near the spiders will quickly push through the door, grab their prey, and pull it back inside where they can feed in the safety of their covered burrow. Some trapdoor spiders will peek out of their burrows and extend a few legs to watch and feel for prey. Others will spin some silk threads that radiate out from the burrow entrance and attach to the burrow's silk lining. These act like trip wires that alert the spider inside to the presence of prey outside. Funnel-web spiders build upon this by having a broad sheet of silk around the entrance, which descends into the burrow. Any unfortunate footsteps on the funnel web can be felt by the spider inside. Some spiders manage to do the same trick using cleverly placed twigs or blades of grass. West Australian trapdoor spiders, *Gaius villosus*, for example, collect long twigs and lie them radially around their circular burrows. The tips of the twigs all point toward the burrow entrance and are attached by silk to the burrow lining. Any large prey disturbing the twigs will be sensed by the spider inside, holding its legs against the silk lining.

European purse-web spiders (*Atypus affinis*) have taken this strategy in a strange direction and rather than just lining their burrow entrance with a tiny smattering of silk, they have extended the silk lining above the ground to form a soft tube. The polite name 'purse-web' doesn't really do it justice; it's more like a flaccid balloon sagging off the top of a bottle, or an old sock hanging off an exhaust pipe, or maybe discarded pantyhose pulled over a mug. This silk purse is camouflaged against the ground by being entirely covered in dirt and debris. So, again, it's less like a purse and more like a discarded dirt sausage. Anyway, most of the time, the wet-sock spider, I mean, the purse-web spider hangs around inside its burrow. When it's time to feed, the spider crawls up inside the dirt sausage and listens for the vibrations of prey walking across the silk. When the prey comes close enough, the spider will pierce straight through the silk with its fangs, envenomating its prey from inside the floppy balloon thing. It will then tear its way through the silk tube, pull its prey inside and then seal the hole in the dirty-balloon-sock-purse-thing.

Trap setters

It's among the more 'modern' araneomorphs that we see spiders building complex silk traps that do much of the hard work in catching prey. You are probably more familiar with these traps than you realise. Every time you have swept away a cobweb from the roof beams or the cellar walls, you have swept away a small spider's purpose-built hunting lodge. Cellar spiders, or daddy-long-legs (family Pholcidae) are some of our most common household spider guests. These delicate spiders sometimes look as if they are floating in mid-air as they sit suspended on a mesh of fine, near-invisible silk. In the corner of a room their webs may be an irregular sheet attached to the walls and floor by a network of fine threads. If you find one of their webs behind a piece of furniture it might look like a three-dimensional mesh of silk threads heading in every possible direction. Any prey that comes too close, or dares try to weave its way through is at risk of being trapped in the dense network of threads. However, even when prey is ensnared in the web there is no guarantee that it will stay there for very long. An insect can often shake or pull itself free from a spiderweb and escape. Web-building spiders can't just sit back and let the web do all the work, they need to quickly immobilise their prey before it can escape. When a spider hears the telltale vibrations of an insect ensnared in its three-dimensional web, it will follow the source of the vibrations and ensure that the insect doesn't escape – by biting its prey and injecting venom, and then wrapping the prey with silk directly from its abdomen. This wrapping behaviour is common in many spiders. Usually, the spider stands on its forelegs, while one pair of hindlegs holds the prey and the other pair of hindlegs pulls silk from its abdomen and places it directly over the prey. When you see this behaviour up close it is hard to track the blur of legs as the spider turns its prey over and over, layering a shimmering silk wrap that leaves the prey looking like a bandage-wrapped mummy.

The loose aggregation of fluffy silk that makes up a daddy-long-legs web seems simple, but it's very effective and it's common to see other spiders fall victim to it. The webs can even be roused into action as a defensive manoeuvre. You may have seen a daddy-long-legs spider in your house suddenly go bananas and shake around in big wobbly circles. This behaviour is called 'whirling' and is used when the spider suspects a predator or a parasite is making its way through the web. By pushing and pulling with its legs, the spider turns its web from a passive trap into a whirling capture net that can turn would-be predators into prey.

At first glance many cobwebs look like a disorganised

mess, but there is often method to the madness. *Latrodectus* spiders, otherwise known as black widows or red-backs, build their webs close to the ground and while they might appear pretty random at first, if you take a closer look you will start to see them take on form. The top of their web is a tangle of threads with seemingly little direction or order to them, but towards the bottom the web consists of more orderly vertical threads that are anchored to the ground. These vertical strands are called 'gumfoot threads' because, after the spider has anchored these threads to the ground, it will coat the base of each one with a layer of sticky glue. These lines are stuck to the ground with just enough strength to hold the silk taut, but they are weak enough to detach if anything comes into contact with them. Any crawling insect unlucky enough to walk underneath this web has wandered into a field of spring-loaded snares. They will need to weave their way through this maze to avoid touching the gooey gumfoot threads. If something like a beetle, say, trips one of these threads with its leg, the thread yanks the beetle into the air by its leg like a bungee cord. The beetle is then thrown into the tangle of silk at the top of the web where the spider is waiting and listening for anything that triggers its trap.

Web builders

We are all familiar with the big round orb-webs that spiders string between trees and along garden pathways, seemingly always at head height. These are made with strong radial threads for absorbing high amounts of energy, combined with sticky spiral threads. With this combination, the webs can simultaneously absorb the impact of a flying insect and trap the intercepted prey in place. Because these webs act as a giant sensory device, web-building spiders generally have very poor eyesight and instead use their webs to listen to the world around them through the vibrations of the silk. A spider sitting in the centre of its orb-web can detect where in the web the prey is, and get an idea of what type of prey and how big it is. The spiders that build these orb webs generally belong to one particular family, Araneidae, and their webs aren't always just rough circles – they come in all manner of shapes and sizes. Ladder-web spiders, for example, weave long vertical webs that are almost rectangular. On one end of the rectangle is the circular 'hub' of the orb web, which then extends either up or down in arching ladder-like threads of silk. Using this type of web, ladder-web spiders can build in places that other orb-web spiders can't. For example, Maude's ladder-web spider (*Telaprocera maudae*) from Australia builds long ladder webs spanning the concave vertical grooves in tall tree trunks. Other orb-web spiders seem to challenge the whole concept of orb webs entirely. *Cyrtophora*, or tent-web spiders, build multilayered horizontal webs. The spider sits under the main orb web, which lies flat underneath a tangle of 'barrier' webs. Any insects intercepted by the barrier webs will ricochet off and tumble down onto the orb-web, like balls in a silk pinball machine.

The spiders that sit out in the open, in the centre of their web, aren't just sitting there for fun; they play an active role in attracting and catching food in their webs. For example, *Argiope* spiders, which are common across the globe, tend to rest with their legs nestled in pairs, giving the impression of an X-shaped spider in the centre of the web. Many *Argiope* have brightly coloured abdomens with contrasting colour patches.

Some have bright yellow bands running across their abdomens while others have shimmering patches of silvery white. Just as the bright colours of crab spiders can pique the interest of insect pollinators, the bright colours of *Argiope* also attract flying insects towards their web, ideally a little too close. This trick even works for nocturnal spiders. The orchard spider, *Leucauge celebesiana*, has a silvery-white and yellow abdomen that attracts nocturnal moths. Jewelled spiders (*Gasteracantha fornicata*) have bold white or yellow bands running laterally across their abdomens. These spiders are usually found sitting diagonally on their webs and it has been shown that, for some strange reason, displaying diagonal lines leads to these spiders catching more flying insects in their webs. Once prey has landed in an orb-web, the spider can rapidly shake the web by pushing and pulling with their legs, which helps entangle the prey in extra layers of sticky spiral thread.

Furthermore, orb-webs aren't just passive traps that sit out in the open and accidentally catch bumbling insects silly enough to fly into them. They are often ingeniously built to lure prey. Some spiders will enhance their webs' attractiveness to prey by building elaborate 'web decorations'. Many *Argiope* weave a large X-shaped decoration out of bright fluffy silk in the centre of their webs. The Australian St Andrew's Cross spider (*Argiope keyserlingi*) gets its name from the bright white X in the centre of its web, reminiscent of the white St Andrew's Cross that emblazons the Scottish flag. From a distance it looks as if these bands of silk extend outwards from the legs of the spider. A number of studies have shown that insects are more likely to fly towards webs decorated with these bright white patterns. The size and shape of web decorations varies immensely; sometimes the X-shaped decorations of *Argiope* will only consist of one to three arms. Some species have decorations that run in a vertical line through the centre of the web, while other orb-web spiders decorate their webs with a fluffy white disc, a radiating spiral, or scattered tufts of white silk. Spiders in the genus *Cyclosa* have been given the delightful common name 'trashline orb-web spiders' because they build their web decorations out of a collection of detritus, dead skin, insect corpses, and egg sacs.

As impressive as all this sounds, it also seems a bit counter-intuitive. Given that spiders make excellent food for birds and bats, it seems a bit dangerous to sit out in the open as a brightly coloured beacon where 'X marks the spot'. Rather than advertising themselves to predators, this display may actually make the web more visible to predators who would like to avoid flying anywhere near it and risk getting covered in super-sticky silk. Alternatively, there's evidence to suggest that web decorations could make the spiders seem bigger and more intimidating, or deflect predator attacks away from the spider's body and towards the expendable silk.

Another family of web builders, the Uloboridae, take a different approach by not relying on silk coated in adhesive; instead, they use a type of silk called cribellate. Up close, cribellate silk looks almost fluffy. It is made of a taut central strand covered in densely folded and bundled soft silk fibres. And while it looks soft and fluffy, it is anything but forgiving. Rather than their silk sticking like glue, it sticks like Velcro. Some Uloboridae weave webs that look like your run-of-the-mill orb-web, others weave what is often simply called a triangle web. The triangle web of *Hyptiotes cavatus* has always been regarded as a little bit clever, but due to a recent discovery is now regarded as a little bit kick-arse. The web looks like a horizontal triangle; on the pointy side waits the spider, standing on a leaf or a branch and holding onto a piece of silk. This thread of silk fans out to where it is anchored to another surface. In this configuration, the pointy end of the web isn't adhered to anything, it is pulled taut by the spider like a kid pulling back a slingshot. This means that it's storing kinetic energy just waiting to be released. When an insect hits the triangle web, the spider releases its grip on the surface and is propelled sideways with the web, which is now loose and compressing itself around the prey. Recently scientists showed that the elastic energy stored by these small webs is completely off the charts. Using high-speed filming, they measured the speed at which the silk and the spider are propelled. Spiders were clocked accelerating at over 770 m/s^2. This is approximately 26 times faster than a space shuttle, or about 54 times faster than a Formula One race car.

Net casters

The Deinopidae family is another group that uses fluffy-looking cribellate silk in wonderfully inventive ways. These spiders are commonly referred to as net-casting spiders or ogre-faced spiders. I choose to use the former name as the latter is unfair for such beautiful beasts. Net casters have two large forward-facing eyes, the largest of any spiders relative to body size. They hunt at night, and these super-sized eyes give them incredible night vision that is more sensitive in the dark than even cat or owl eyes. Unlike other web builders that weave webs out in the open, net casters take matters into their own claws by building their webs in between their legs. At night, these spiders hang upside down somewhere near the ground. They lie still, holding their small rectangular webs between their four front legs, and wait. Using their night-vision eyes they can see any insects walking around on the ground beneath them. When an unlucky insect wanders too close, the net-casting spider will lunge forward and spread out its web. In one movement the spider will scoop its prey into the web, like a hand-thrown fishing net.

Even though they have fantastic night vision, net-casting spiders aren't limited to catching prey they see on the ground. Recent research has shown that North American *Deinopis spinosa* can also hear flying prey using small sensory organs on their legs. When a fly buzzes behind their heads, net casters can hear them coming and will quickly sweep their net backwards, snatching the fly out of mid-air.

Glue spitters

At some point in your life, you have probably seen a sci-fi or horror movie where a giant spider-like creature traps their helpless victims by spitting or spraying them with silk. If not that, then you've probably seen a version of Spider-Man swinging through the streets shooting webs out of his wrists. This is a bit of a fantasy bubble that needs bursting. In reality, spider silk glands do not have any muscular or vascular mechanism to push or project silk out of them. Silk must be pulled out by some force – like the spider pulling the threads with its own legs, or a spider hanging from a piece of silk and having the weight of its own body draw the silk from its abdomen. You can draw silk directly from a spider yourself if you like. With some careful handling you can use a pair of forceps, to pinch a thread of silk near a spider's spinnerets and slowly reel it out of them. The more you gently pull, the more silk the spider makes. If this sounds a bit silly, heed this warning: there are some ludicrous silk-milking stories coming up soon.

As is becoming a theme in this chapter, there are exceptions that prove every rule, and the aptly named 'spitting spiders' bend the rules just enough to give some credence to the idea of spraying sticky tendrils to catch prey. *Scytodes* spiders sit on a small web, often woven across the surface of a large leaf. The web isn't sticky as it is mainly there to help the spiders detect their prey. They don't have great eyesight so they sit 'listening' with their legs for any vibrations. Once they detect prey nearby, they will turn to face it and spit streams of a thick glue straight out of their fangs. Shaking their fangs side-to-side, they spit a zigzag mesh that begins drying on contact, holding the prey in place. To finish the job, the spitting spider runs over to its glue-covered prey, injects it with venom and wraps it tightly with silk. Their glue is made in modified venom glands and chemical analyses of the glue from *Scytodes thoracica* has shown that it's part adhesive, part venom, *and* contains proteins that share some interesting similarities with spider silk proteins. If you look at the strands of glue under a high-powered microscope it appears to have a silky-looking fibre running through it. So, maybe the idea of sci-fi silk-spitting spiders isn't so far-fetched after all.

These are just a few of the incredible ways that spiders can use silk and webs to catch prey. There are seemingly endless modifications on these techniques, depending on what type of prey the spiders are trying to catch. There are even 'kleptoparasitic' spiders who will sneak onto other spiders' webs to steal pre-captured food. This arsenal of techniques is made possible by the miraculous material that is spider silk. Later, we have a whole chapter devoted to the beauty and complexity of silk. Now, it's time we move on to the second major weapon spiders have at their disposal – venom.

What is spider venom?

Almost every spider is venomous in some capacity. Once again, there's always that annoying exception to every rule and, in this case, it's the Uloboridae. As well as having soft, fluffy, and sometimes super-accelerating webs, the Uloboridae family are exceptional because they have completely lost their venom glands. They rely entirely on silk to catch and subdue their prey. If, for example, a small insect becomes tangled in their Velcro-like webs, the spider will finish the job by pouncing on the insect and wrapping it tight with even more silk. With the prey securely immobilised, the spider can simply walk over to it and start regurgitating digestive fluids straight onto it. And they don't just start digesting one part of the prey, they cover the entire thing in digestive fluids while it's still alive, trapped in thick compressive silk. Why these spiders lost their venom glands is a bit of a mystery and, if you ask me, it all sounds a little bit sadistic. Thankfully, every other spider on Earth is merciful enough to put prey out of its misery with a quick shot of venom.

Unlike the trappers and web builders above, many spiders don't bother using any silk to catch prey and instead rely on stealth, strength and potent venom. Some jumping spiders, wolf spiders, funnel-webs, tarantulas, and many other spider species, will catch prey simply by jumping on top of it, holding on for dear life and injecting venom as quickly as possible. For strategies like this to work, venom needs to be delivered directly into the prey and take effect as quickly as possible.

Spiders have dedicated glands near their heads that produce venom and lead directly to the fangs. Fangs sit at the tip of a small pair of appendages called the chelicerae, which are right under the spider's eyes. Venom can be incredibly complex. One species' venom can contain thousands of different chemicals. A handful of species have had their venom studied, but for most we don't know exactly what's in it, or how it works. Most of the active venom chemicals we understand are neurotoxins. They are injected inside a spider's prey and start damaging nerve cells, messing with the central nervous system. This can paralyse prey by stopping nerve signals from passing through or, rather disturbingly, stop nerve cells from being able to turn off, making the prey's nerves uncontrollably fire and spasm until they finally die. Many spider venoms are made of peptides, which are very small molecules consisting of only a few amino acids. But these small molecules can pack a big punch. They are like a spanner in the works of the biological processes going on inside animal bodies. They get inside and grind chemical reactions to a halt. Each spider species has its own special cocktail of venom chemicals that have evolved to tackle different types of prey. Different species' venoms interrupt different types of biological processes. This is why spider venom can affect one type of animal but not another. The biological processes going on inside an insect, for example, are very different to those going on inside a mammal. So, a particular venom peptide can be lethal for an insect while having very little effect on a mammal.

Once a spider has bitten its prey and the venom has done its job, the next step is feeding and digestion. Some spiders will wrap their prey in silk and hold onto it for later; others will start eating right away. When the spider is ready to eat, it will begin by regurgitating digestive fluids onto its prey. So, the digestive process actually begins outside of the spider's body. The fangs are mostly there to deliver venom; they aren't really involved in the feeding process. Some spiders, like large mygalomorphs, will use their fangs along with a few other appendages to mush their prey a bit, making sure the digestive fluids get sloshed around inside to make everything nice and pulpy. But while a feeding spider might look like it is eating prey with its fangs, or sucking the innards out of its prey, the truth is a bit sloppier. Instead, the digestive fluids are simply placed directly on top of the prey, enzymes in the fluid start turning the prey into a liquid mush, and this nutritious mush is then sucked back up into the spider's mouth. Essentially, they just vomit on their food, wait a bit, then slurp the vomit back up again. Simply majestic creatures, right?

How venomous are spiders?

Asking how venomous a spider is, is a weird question. Venom potency doesn't come down to something as simple as how concentrated it is, or how much a spider can inject. There are many different types of venoms that work in different ways, and they have different effects on different types of animals. So, there is no way to objectively rank spiders from least venomous to most venomous. But let's face it, when we talk about how venomous spiders are, we're not really interested in how they use it to catch prey – what we're really interested in is whether they can kill us. The real question is: how venomous are spiders to humans?

The short and boring answer is, not very. Since spider venoms have evolved primarily to take down small invertebrates, their venom doesn't have much impact on us. And when it does it's probably just a weird quirk of biochemistry, not some evolutionary history between spiders and humans. Spider venom is primarily for prey capture rather than defence against large animals. There has been extensive research into the toxicity of spiders and the medical conditions arising from spider bites, and the general conclusion is that all but a few species are of no real medical concern. Most spider bites affect people like bee stings do; there can be local pain and swelling, but that's about it. The usual treatments for bites are straightforward things like rest, cold compression and pain relief. As is the case with any bite or sting from an animal, there are rare cases of a severe allergic reaction, and the very young and very old can suffer more complications from spider bites. But overall, medical research points to most spider bites being low impact conditions that can be monitored and treated by people at home.

Even big scary looking tarantulas with giant fangs aren't of any serious medical concern. Many people keep tarantulas as pets and will quite happily let them crawl all over themselves without any fear of being bitten. Their bites, at worst, can be painful simply because of the size of their fangs, and the effect of their venom has been likened to a bee sting. You're at a greater risk of getting tarantula hairs in your eyes than you are of getting a serious bite. Tarantulas from the Americas have special hairs called urticating setae. These are small hairs with pointy barbs running down their length that are there to deter predators. There are numerous medical reports of people handling pet tarantulas only to find their skin irritated from the small hairs or, even worse, they rub their eyes and get the hairs stuck in their corneas.

But you're not reading this part of the book to hear that rational nonsense, are you? You're here to get the dirty gossip on the spiders that can really mess us up. Alright, let's dive in. Medical research points to a small percentage of spider species whose bites can inflict more than just some localised pain and swelling. In rare cases these bites can be life threatening. Let's go through the nasty ones here and finish by talking about the spider that I know you are here to read about – the world's most venomous spider: the Sydney funnel-web. But first, let's start with the beautiful black widow spiders.

Latrodectus: Black widows and red-backs

In the Americas they're known as black widows; in Australia, red-backs; in New Zealand they're katipō spiders. Whatever you call them, they are easily identifiable spiders in the genus *Latrodectus*. They are small and glossy-black, often with a bright red pattern on their abdomens. The Australian red-back (*L. hasselti*) has a jagged red strip running front to back. The South American black widow (*L. curacaviensis*) has thick bands of red running across its back, while the North American *L. mactans* has the iconic red hourglass shape on its underside. African *Latrodectus* can be a lighter brown with white stripes. There are over 30 species of *Latrodectus* all over the world and they are all considered venomous to humans.

They like to live in dark, dry places like garden sheds and under patio furniture. The kinds of nooks and crannies that people love shoving their hands into without looking first. In the days of outdoor toilets, people's buttocks and genitals were prime targets for red-back fangs. Their venom contains neurotoxins that affect your nervous system. This is the part of the book where you might be expecting me to go into all the gruesome details about people dying horrific deaths – breathing their last breath and watching the world turn dark as they stare wide-eyed at the black and red horror poised on their arm. Sadly, I must temper your expectations. This chapter will be a bit more boring than that. Most *Latrodectus* bites are pretty low impact and result in local pain and swelling. In studies of North American black widows its estimated that around half of all bites are 'dry bites' – bites where the fangs don't actually inject venom. A small portion of confirmed bites result in severe envenomation where symptoms can include spreading pain, high blood pressure, fever and muscle spasms. At worst these can last a few days and for severe cases there are antivenom drugs available for treating *Latrodectus* bites.

Phoneutria: Brazilian wandering spiders

Phoneutria aren't restricted to Brazil, and can be found throughout the northern end of South America and parts of Central America. They are sometimes called armed spiders and when they do go wandering it's usually at night time. During the day they are hidden in burrows and crevices. There are only a handful of species in this group and their venom also contains neurotoxins that affect human nervous systems.

When disturbed, they display a threatening behaviour where they rear up on their hind legs, showing off their shiny fangs. This action is unusual in spiders and suggests a rare example of venom being used defensively as well as for prey capture. It's also rather nice of them to give a bit of a warning before they do bite. Less than one per cent of bites result in severe symptoms, and less than ten per cent show moderate symptoms. Severe symptoms can include pain and heart problems that are potentially life threatening in very young children and the elderly. One surprising spider bite symptom that sometimes rears its ugly head is prolonged and painful erections. I know what you're thinking fellas, don't try this at home. But hold that thought, there will be more on this arousing topic later.

Sicariidae: The recluse spiders

The genus *Loxosceles*, in the family Sicariidae, has around 100 species globally, and a handful of them are considered venomous to humans. They have various common names like brown spiders, violin spiders and fiddleback spiders, but the one that most people are referring to when they talk about venomous recluse spiders is the common American brown recluse spider (*L. reclusa*). Recluse spider venom has one particular effect that stands out above all others and strikes an inordinate amount of fear into people. Along with the usual pain and swelling, recluse spider bites can lead to necrosis, which means the skin cells around the bite start to die off. In most cases this is limited to a small red lesion on your skin that can heal itself. In severe cases these manifest as necrotic ulcers that may need to be surgically removed. Even though this is rare, tales of this particular symptom have gone on to have a bigger impact than they should. Necrotic ulcers can be caused by many medical issues but are often misdiagnosed as spider bites, which can just lead to more pain and suffering for the patient who doesn't get a proper diagnosis.

What causes necrosis in these spider bites is now well understood. Recluse spider venom contains proteins with the catchy name of phospholipase D toxins. For now, let's call them PLDs. PLDs work by breaking down phospholipids, which happen to be one of the main chemicals that make up our cell membranes. So, when the PLD toxins from recluse spider venom come into contact with our cells, they start to burst open. PLDs are actually produced as a normal part of cell function in many different living things, but in the wrong context and concentration can do some real damage. Recluse spiders belong to the family Sicariidae along with the gloriously named six-eyed sand spiders in the genera *Hexophthalma* and *Sicarius*. Both of these genera have now also been shown to have PLD proteins in their venom and can cause necrotism. No other spider bites have necrotic effects. The idea that spiders could have bacteria in their venom or on their fangs that could also cause necrotism has been debunked. Sadly, 'flesh-eating' wounds have been erroneously blamed on all sorts of different spider species all over the world that are completely incapable of inflicting this kind of bite. The following chapter is entirely dedicated to exploring this and other common spider myths. So, before you go telling your friends about necrotising spider bites, read the next chapter to make sure you aren't spreading any more fake spider news.

Hexathelidae: Funnel-web spiders

This is it. This is the big one. The Sydney funnel-web (*Atrax robustus*) holds the title for world's most venomous spider. Only people in south-eastern Australia need to worry about these ones. And even though they live in Australia's most populous city, they are rarely encountered. They spend most of their time minding their own business inside their funnel-shaped burrows. Most bites occur during mating season in late summer when male funnel-webs leave their burrows at night in search of females and end up seeking shelter in places they shouldn't, like people's discarded shoes.

There are two other species of *Atrax* funnel-webs and some more from the closely related *Hadronyche* and *Illawarra* genera. They are also found in south-eastern Australia and have similar venom that can deliver a serious bite. It's unclear whether any one species of funnel-web has more potent venom than the others, but the Sydney funnel-web holds the 'most venomous' title because there are more reported bites from this species than the others. This is probably because it lives in urban areas and people are more likely to come across them than the other species, which live in more remote areas. *Illawarra* funnel-webs, for example, are only known to be highly venomous from research into the chemistry of their venom; there aren't actually any recorded bites from these species to confirm their highly-venomous status, and fingers crossed there won't be any.

Funnel-web spiders are the only large mygalomorphs on this list of potentially venomous spiders. Like wandering spiders, funnel-webs are famous for their defence behaviour of rearing backwards, raising their front legs and readying their fangs. It's a good thing they give a bit of a warning, because their bites are nasty. One of the characteristic symptoms of a funnel-web bite is immediate severe pain. This, plus large, bleeding fang marks, leaves no doubt as to whether or not you've been bitten by a funnel-web. They have neurotoxic venom and can deliver large amounts of it with a single bite. Severe effects can occur in around 13 per cent of bites, and these include nausea, vomiting, headaches, muscle spasms, numbness, fluid build-up in the lungs, and cardiovascular and nervous system effects. Unlike other spider bites that have symptoms which progress slowly, a severe funnel-web bite can fully escalate in under an hour. Before the development of antivenom, deaths were reported to have occurred within the hour. Since symptoms can manifest so quickly, it's best to seek medical advice immediately even in the case of a suspected bite. Thankfully, as with other venomous spiders, most bites result in mild to moderate symptoms. Antivenom is available in Australian hospitals and can completely reverse the symptoms of even severe bites within a matter of days.

Other potentially dangerous spiders

The genera *Latrodectus* and *Phoneutria*, and the families Sicariidae and Hexathelidae are the big four groups of spiders that are considered dangerous by medical professionals and venom researchers. That's not to say that other spiders won't give you a painful chomp, but they probably won't put you in a hospital bed. Since spider bites are so rare it's hard to get good data on them, and we are learning more about weird and wonderful spider bites every day. The Indian tarantula genus *Poecilotheria* are now known to give bites that cause painful muscle cramps for a few days, which is unlike any other species of tarantula. And there are probably other spiders whose venom is harmful to humans, but the chances of them ever coming into contact with people, not to mention biting people, are so low that it just never happens. Australian mouse spiders (*Missulena*), for example, are closely related to funnel-webs and have similar venom. There's anecdotal evidence that their bites are quite virulent, but bites are so rare that good data on their medical effects simply doesn't exist. Experts are always keeping an eye on whether other spiders should have their venomous status upgraded. As the old saying goes, it's better to be safe than sorry.

You may have heard stories of horrendous bites from spiders that aren't on this list. You might have met someone who swears hand-on-heart that some other species of spider put them in a coma for three years until they were finally woken up by the scent of their favourite 40-year-old triple-distilled whisky. Some of these stories are pure myths, and we'll deal with those in the next chapter, but many of them are just a case of mistaken identity. It turns out that properly diagnosing a dangerous spider bite is a lot harder than it seems.

Diagnosing and treating spider bites

I should make one thing very clear here; while I sometimes go by the title of Dr James, I am not that kind of doctor. So, don't take anything in this book as medical advice. And please don't send me pictures of your suspected spider bites. I can't help you and I am a little bit squeamish. Diagnosing and treating spider bites isn't just a run-of-the-mill medical problem. Because of the relationship we have with spiders, understanding their bites is part medicine, part zoology and part psychology, with sprinkles of mythology and sociology thrown in for good measure. Doctors and nurses can't be expected to be trained arachnologists and toxicologists too, so mistakes are common when diagnosing spider bites. For this reason, experts have put forward three criteria that must be met for a condition to be conclusively diagnosed as a spider bite:

1 **The spider bite must be witnessed:** This includes seeing and/or feeling the spider bite at the time it actually occurs. This avoids issues around people who find strange wounds or puncture marks but don't remember being bitten. They suspect it could be a spider bite, but in actuality it could be anything from a bee sting, to a prick from a rosebush, to a fungal rash.
2 **The spider species needs to be identified:** Ideally this involves the spider that bit someone being immediately collected and taken to be identified by an expert arachnologist. For many people, including medical professionals, all spiders can look alike and bites are commonly blamed on the wrong species when a proper identification isn't made. This criterion also avoids problems where people may feel or see a bite happen, and then just blame any old spider that they have seen around the house that makes a convenient scapegoat, instead of the actual spider that bit them.
3 **There should be clinical effects directly linked to the bite:** These can include pain, redness, swelling, or the presence of puncture marks in the skin. With so many people suffering from spider fears and phobias, people have been known to present at hospitals suspecting spider bites that didn't actually occur. If someone sees a spider

on their arm, they might jump to the conclusion that they have been bitten when they haven't.

So, if you do think you have been bitten, resist the urge to flatten the spider beyond recognition with a rolled-up newspaper. Collect it in a sealed container and, should you need to seek medical treatment, the intact spider will be immensely helpful for properly diagnosing the bite with an accurate species identification.

Again, most spider bite symptoms are mild to moderate. And in those unfortunate cases where the effects are severe, we are lucky enough to live in the age of antivenoms. There are antivenom medications available for many spiders and they are incredibly effective. Even for the world's most venomous spider, there hasn't been a single death resulting from a Sydney funnel-web since the invention of funnel-web antivenoms in the 1980s. This doesn't mean that you should be flippant about spider bites. Antivenom is used sparingly and only in severe cases. You might be wondering, *if antivenom is so great, why isn't it used more often? Why aren't Sydney-siders walking around with vials of antivenom in their bags, next to their sunscreen and pawpaw ointment, just in case they get bitten by a funnel-web?* It turns out that antivenom is really expensive. And the way it is made is very laborious and, quite frankly, a bit gross.

I grew up in Sydney and every summer I would hear familiar radio stories and read familiar newspaper articles about funnel-web spiders being active again. These stories always came with a reminder for people to do sensible things like shake out any boots they had left outside on the off-chance a funnel-web wandered into them. There would also be a reminder that if you do happen to find a funnel-web, you shouldn't kill it or even just ignore it. You should carefully collect it and deliver it somewhere, like a local zoo or museum, where its venom can be milked and used for making antivenom. I never really thought about this very much. I guess I just assumed that the spider venom would be put into a vial and brought to a lab where they looked at it under a microscope and did some chemistry stuff with beakers and pipettes and, voila, you get some antivenom. When I finally found out how life-saving antivenom is produced it made me feel all sorts of conflicting emotions.

To make antivenom, first you need a lot of spiders. Their venom can be 'milked' by someone expertly handling them and getting them to express venom. Or, to put it another way, they annoy the spiders enough that they try and bite, and then quickly suck up venom droplets from the tips of their fangs with a pipette. As annoying as it must be, this doesn't hurt the spider and they can be returned to the wild afterwards.

Once you have milked a whole bunch of spiders, the next thing you need is a whole bunch of rabbits. Rabbit immune systems are very good at making antibodies that neutralise venom molecules. So, if you inject a rabbit with a small amount of spider venom their immune system will start making spider venom antibodies. Then you gradually increase the amount of spider venom injected into the rabbits and ramp up their immune systems' antibody production. Do this with a whole bunch of them and you will soon have a population of rabbits supercharged with spider antivenom that can be extracted from their blood. You purify the antibodies from the rabbit blood and shove it into a human. Simple, right? The process is similar for other types of antivenom, such as for snake bites. The main difference is that for snake antivenom you use horses instead of rabbits.

At the end of the day, the best approach is to avoid getting bitten at all. As is the case with most wild animals, leave spiders alone and they will leave you alone. There isn't a single spider in the world that goes out looking for people to bite. Most bites are unfortunate accidents where the spider is disturbed and has no other way of escaping, such as when people are gardening or they put on a shoe or piece of clothing with a spider inside. Overall, spiders just want to be left alone and will avoid people if they can, only biting as a last resort. You have to either be very unlucky, or very silly to get bitten by a spider. So, if not for your own health, for the sake of those poor rabbits being pumped full of venom, please don't get bitten by a spider.

And while antivenom is great, I would argue that it is perhaps the least interesting medical use of spider venom. The chemicals inside spider venom are being teased apart for their use in developing new drugs. They have also been touted as the next generation of environmentally friendly pesticides for agriculture. And with so many spider species, each with its own unique venom cocktail, spiders are a goldmine for bioprospecting new wonder drugs.

Spider venom in medicine

We've spent a lot of this chapter talking about how great spiders are at eating other things. But let's take a moment to remember that more often than not, spiders are a tasty meal for many other animals. There is a species of small bat (*Myotis emarginatus*) that actually specialises in hunting spiders. At night these bats will use echolocation to seek out web-building spiders and pluck them right out of their webs, with spiders making up around 75 per cent of the bats' diet. Even humans are partial to snacking on a spider or two. Large tarantulas like the Goliath birdeater in South America are eaten as a traditional bush food. Tarantulas and huntsman spiders are commonly prepared in food markets in China and South-East Asia. The golden orb-web spider *Trichonephila edulis*, common in the South Pacific, derives its scientific name from its edibility – *edulis* is Latin for edible.

In Cambodia, the town of Skun has become world famous and is affectionately called 'spider town' for the local practice of cooking and eating tarantulas. Legend has it that the town was forced to eat spiders during the widespread famines of the Pol Pot regime and it has since become tradition. There isn't a whole lot of evidence for this tale. Nevertheless, the Skun marketplace has become a tourist hotspot for those eager to try a deep-fried spider, or those who just wish to see the stalls packed high with this delicacy.

Spiders also feature in traditional medicine, with claims that spiders are good for pain relief, respiratory problems, and even as aphrodisiacs. Researchers have described how the Chol Mayan people of south-eastern Mexico use the red rump tarantula (*Tliltocatl vagans*) in traditional medicine. Medicine men in these communities conduct a special ritual when preparing a mixture of crushed tarantula, alcohol, tobacco and garlic. This mixture is filtered and then drunk by the patient to help respiratory conditions like chest pain and coughing.

Modern medicine also has its sights set on spiders in the hope that they hold a key to new pharmaceuticals. As mentioned before, spider venoms are a bit like chemical spanners dropped into the moving gears of biochemical processes. If we extend this rather simplified metaphor a bit, we can start to see how venoms can actually be used as a helpful tool in medicine.

Disease is just another biological process that happens inside our bodies; it's just a process that acts against us. It turns out that some spider venom molecules can be used to grind these processes to a halt for our benefit. The lethal effects of spider venoms are being investigated for their application in new types of antibacterial, antiviral and antifungal medicines. And their cell-busting abilities hold some exciting potential for being able to attack cancer cells. Laboratory studies have identified a number of spider venom compounds that assail certain types of cancer cells, while having little effect on healthy cells. Toxins from tarantula venom have been shown to stop the spread of brain tumour cells. Asian funnel-web toxins can stop the spread of breast cancer cells, and wolf spider toxins can stop the spread of lung, colon, prostate, cervical and liver cancer cells. Great caution is being taken in developing these into drugs for human use. They are toxins after all, and any unwanted side effects need to be understood before human trials can go ahead.

Since many spider toxins can have a paralysing effect on nerve cells, this has led scientists to question whether this feature can be used to interrupt specific nerve impulses that we don't like, such as chronic pain. Certain drugs currently used to treat chronic pain, such as opioids, can be addictive or lead to the development of drug resistance. Animal venoms are being intensely studied to see whether they might hold the secrets to a safer pain relief alternative. Certain pain signals come from special nerve cells called nociceptors. If there are spider venom toxins that can affect the function of these nerve cells, while having little impact on other essential nervous system functions, then they could be good candidates for venom-based pain killers. Obviously, these should be the venom compounds that have a paralysing effect, as opposed to the many venom compounds that have an excitatory effect on nerve cells, which could cause intense pain rather than cure it. Currently, one of the hopefuls for delivering these drugs is the Brazilian wandering spider *Phoneutria nigriventer.* Its venom has been studied extensively and a number of different venom compounds have been identified. Quite a few of these have a paralysing effect on pain sensors. Their varying effects and side effects are being studied to test whether the venom compounds could be used as treatments for different types of pain related to cancers, postoperative symptoms, and damaged nerve tissue.

Also, remember above we mentioned that some Brazilian wandering spider bites can lead to prolonged and painful erections? It doesn't take a genius to realise why the pharmaceutical industry is very interested in this. Scientists are trying to harness the erection-inducing qualities of wandering spider venom to develop new treatments for erectile dysfunction. Erectile problems often stem from vascular issues that affect the flow of blood into erectile tissues. Studies in lab rats have shown that a compound from wandering spider venom can lead to relaxation of erectile tissues that helps with blood flow into these areas. With a bit of chemical wizardry, scientists have been able to modify the venom peptide so that it can be used to increase blood pressure in lab rat erectile tissues without the toxic effects of the spider venom.

These are just a few examples of how spider venoms are being studied for use in human medicine. Venom peptides also hold promise as potential treatments for heart disease, stroke, epilepsy and muscular dystrophy. Applications aren't limited to human medicine either. Spider venom may be the key to treating things like Tasmanian devil facial tumour disease.

When Stuart Harris took this photo of an as yet undescribed *Maratus harrisi* he had no idea that it would change his life

Male *Maratus elephans* peacock spider displaying

Male *Maratus fimbriatus* displaying

Female *Myrmarachne smaragdina* spiders from northern Australia make convincing mimics of green tree ants

Orsima ichneumon jumping spiders are said to be backwards-mimics of ants

A *Thomisius* crab spider hiding inside an *Asystasia* flower in the Malaysian rainforest

An Australian wolf spider exploring at night. The fangs sit at the tip of the chelicerae, which are directly below the forward-facing eyes. Either side of the chelicerae are the slender pedipalps.

A female *Argiope aemula* with prey in her web, with white silk 'web decorations'

An Australian net-casting spider (*Asianopis subrufa*)
holding its cribellate prey capture web at the ready

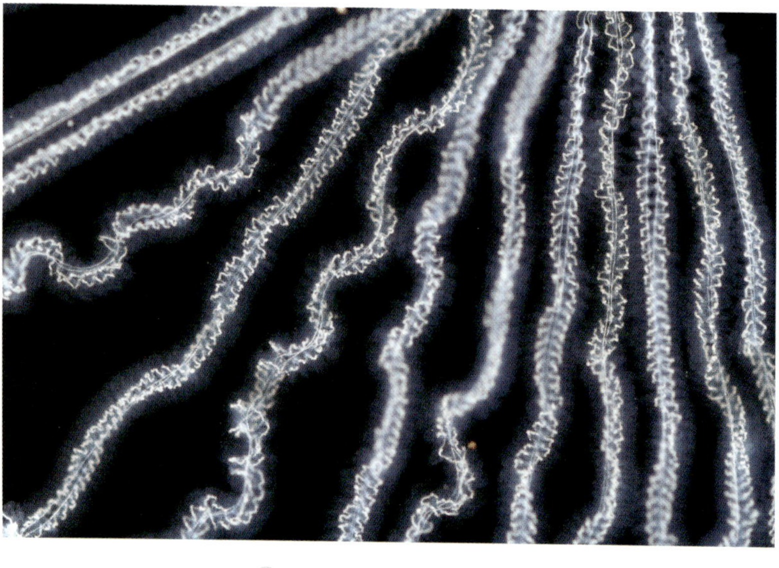

A close-up view of the fluffy cribellate silk of a net-casting spider

Cape made entirely from woven
golden orb-web spider silk

Adult female *Nephila pilipes* in Malaysia, showing the golden silk that is characteristic of golden orb-web spiders. This species' silk has been used to weave playable violin strings.

A large Australian *Trichonephila edulis* female distracted by her prey, while a smaller male sneaks in to copulate

The barely visible scorpion-tailed *Arachnura higginsi* sits at the end of a long string of silk egg cases suspended in her web

This huntsman mother stands guard over her egg case even after it has been damaged in a pile of dry firewood

A female fen raft spider (*Dolomedes plantarius*) guarding her nursery web

An Australian garden orb-web spider (*Hortophora transmarina*). Eight of these spiders were sent into space on board the space shuttle *Columbia*.

A golden orb-web spider (*Trichonephila clavipes*) and the web that it built in microgravity on board the International Space Station.

Tasmanian devils are an endangered species in Australia and their populations are being rapidly killed off by devil facial tumour disease (DFTD), a severe and highly contagious type of cancer. There are few effective treatments for DFTD and it's estimated that without intervention Tasmanian devils could be extinct within about 20 years. Scientists have discovered that a particular peptide called gomesin, found in some tarantula and funnel-web spider venoms, targets DFTD cells. The peptides kill the cancer cells and stop their spread without affecting the animals' healthy cells. This treatment, alongside things like captive breeding programs and reintroductions, could help us win the fight against DFTD and save the Tasmanian devil.

While all this research is very promising, we are still a few years away from spider venom peptides being sold at the pharmacy. There are a few currently going through human clinical trials, so watch this space. There is, however, another use for spider venoms in agriculture that has already found its way into the market and is being widely used right now.

Spider venom in agriculture

Since spider venom has evolved over millennia to kill small invertebrates, it makes sense that these chemicals hold promise as candidates for new organic insecticides. Earlier I mentioned that spider venoms are often neurotoxins, which target a prey's nerve cells. This is actually pretty similar to the way many common insecticides work. For example, many fly sprays work by binding to a chemical called acetylcholinesterase. This chemical is what the body uses to turn nerve impulses off, so by stopping acetylcholinesterase from doing its job, the fly spray keeps the fly's nerves firing uncontrollably until it spasms itself to death. This is why you might have seen a fly hit by insecticide madly buzzing around on its back before slowly dying a torturous death. Just something to think about next time you reach for the can of fly spray.

One big problem with these types of insecticides is that chemicals like acetylcholinesterase are common in many different animals. So, insecticides that use this type of method are broad spectrum and can have many unintended consequences. Ideally, large-scale agricultural pesticides should target the pest animals without having any impact on the other wild creatures living nearby. This is where spider venoms can help. With so many different types of venom that can target different types of animals, scientists are trying to find venom molecules that can target pest insects without any collateral damage. Add to that the fact that spider venom compounds are organic and biodegradeable, and thus shouldn't impact soil and water quality like some synthetic pesticides.

The US-based pesticide manufacturer Vestaron boasts that its 'Spear' line of products only targets insect nervous systems, without affecting mammals, birds and fish. The active ingredient in these pesticides is a neurotoxin derived from the Australian Blue Mountains funnel-web (*Hadronyche versuta*). It's the first product of its kind on the market and there are hopes that future pesticides can improve on its selectivity. A toxin from the American desert bush spider (*Diguetia canities*) has been shown to kill the highly invasive German cockroach without having any impact on native American cockroaches. Another toxin, from the African tarantula *Augacephalus ezendami*, has been shown to kill fruit flies and paralyse sheep blowflies, but it has no impact on moth larvae.

One of the hurdles with developing spider venoms into pesticides is the fact that they are usually injected directly into insects with fangs. Many venom chemicals will have no effect if they are sprayed on, inhaled, or ingested by the pest. Any commercial pesticide derived from spider venom will need to pass this test unless we plan on sending farmers out into their crops with billions of miniature syringes. You might be thinking, *if spider venom is so good at killing insects, why not just let a bunch of spiders loose in crops?* That's actually a pretty good idea, and we'll get to that later in the book. For now, the take-home message is that with advances in medical research, spider venom can help us much more than it can harm us. Sadly though, spiders and their venom are the source of endless myths, misinformation and fake news. So, let's take some time to clear the air, and get to the bottom of some of the most common spider myths.

Chapter 5

MYTHS AND MISCONCEPTIONS

The sight of a palm-sized huntsman on the kitchen wall is as familiar to Australians as the smell of fresh sunscreen on a hot Christmas Day. These large brown spiders in the family Sparassidae are common household visitors. One of my favourite Aussie pastimes is enjoying the thrill of having a huntsman scuttle across your car windscreen while you're roaring down the highway at 100 kilometres an hour, concentrating on driving in a straight line, and waving your hand around trying to find the windscreen wiper switch to figure out whether the spider is on the inside or the outside. This iconic spider is as Australian as wearing a cork hat and thongs. There was even a song written about them: 'Football, meat pies, kangaroos and huntsman spiders!' At least, I think that's how it goes ...

Though the more I think about it, huntsman spiders aren't just Australian; they belong to the world. If this book were written by anyone else, this chapter might mention that huntsman spiders are iconic for Americans, Thais, Moroccans or Spaniards. There are over 1000 species of Sparassidae across the globe. The world's largest spider, by leg span, is a giant huntsman from Laos (*Heteropoda maxima*). The world's fastest spider is the Moroccan flic-flac spider (*Cebrennus rechenbergi*), which cartwheels up and down sand dunes at speeds of up to two metres per second. What makes huntsman spiders so familiar to us all is their knack for making homes inside our own. Huntsman spiders have flattened bodies that let them crawl under bark, through crevices in rocks, and in small gaps under windows and doors. They manage to find their way inside our homes and even our cars, where they will happily make a living feeding on small insects that also find their way indoors. People across the globe are united in their close relationship to huntsman spiders. Almost everyone you speak to will have an anecdote of some sort about the huntsman spiders they have come across in their day-to-day lives.

I had a funny encounter with a huntsman spider when I was about 16 and woke up in the middle of the night with one on my face. I was stirred from my sleep and realised that I could feel the weight of something pressing against my cheek. Instinctively, I swiped at whatever it was and sent the poor creature flying across the room. In the darkness, and in a half-awake daze, I could just make out the silhouette of a large many-legged creature soaring off into the distance. For some reason my tired and confused brain jumped to the conclusion that it was a frog, and that I had slapped a wet frog across the room. I lay awake confused and anxious about why there were frogs in my room and what they were doing on my face. Eventually the adrenalin subsided and I realised what it actually was. It made much more sense, given huntsman spiders were pretty common visitors to the house and frogs generally don't have that many legs. I fell asleep again, giggling to myself about the idea of a frog splattered somewhere on one of my bedroom walls.

Later, when I told people my story, some found it funny,

others found it scary, but a few seemed to think that the story made complete sense. They told me that spiders are known to crawl into your mouth while you sleep and drink your saliva, and that must have been what the huntsman was up to. This sounded utterly bizarre but I came to learn that this is a widely circulated myth. To be clear, there is not a shred of evidence to back this up. Trust me, if there was, arachnologists would be excitedly setting up night-vision cameras to assay the biodiversity of spiders aggregating around their own spit puddles. And I mean, come on mate, how arrogant would you have to be to believe this kind of myth? No-one, not even a spider, is thirsty for a slurp of your stinky face broth in the middle of the night.

Like many myths, we probably have the internet to thank for the proliferation of this one. And there are variations on it as well, like the idea that spiders will drink the fluid from your eyes while you sleep. Again, to be abundantly clear, this is bollocks. Another common myth that is probably linked to this one is the idea that we swallow spiders in our sleep. I see this one pop up every now and again on the internet, with supposed statistics to give the myth an air of validity. The common false statistics seem to be that we swallow up to eight spiders a year, or an average of 52 in a lifetime. Again, bollocks. There is not a single shred of evidence to give any credence to this bananas idea. My story about waking up with a spider on my cheek may very well be the closest any human has ever come to eating a spider in their sleep. And, even then, there's still a chance it wasn't even a spider! Maybe it was a frog after all!

Spiders are the unfortunate victims of many myths, misconceptions and outright lies. So, while we're here, let's clear the air and debunk some of the most common ones. Starting with the myth of the necrotising, gangrenous, white-tailed spider.

Four common spider myths debunked

White-tailed spiders *do not* have necrotising venom

For me, this particular myth stands out above the others because it is something that I grew up hearing repeated over and over again. When something is repeated so often and so confidently, you might assume that it must be a fact. Thankfully, this one is a complete furphy.

White-tailed spiders are very common household spiders in eastern Australia. There are actually two species (*Lampona cylindrata* and *L. murina*) that look very similar; they are small and brown with a bright white patch at the very tip of their abdomen, and are apparently happy living in cupboards and wardrobes. You may have heard, like I did, that their bites cause 'necrotising ulcers' on your skin. The idea at its worst was that there was something in their venom, perhaps a kind of flesh-eating bacteria, that caused your cells to start dying off around the site of the bite. This necrosis would start to spread and eventually lead to a great big gaping hole-of-death in your skin that would need to be cut out before your limb turned gangrenous and the whole thing needed to be amputated. Again, just to labour the point, this is false. The idea seems to have appeared somewhere around the mid-80s when doctors began misdiagnosing skin necroses as coming from spider bites. The idea took hold, more doctors started misdiagnosing the condition, the idea spread through the media, and it snowballed from there. White-tailed spiders were accidentally introduced to New Zealand from Australia somewhere around the early 1900s. The myth also managed to jump borders and from the mid-80s onwards, false diagnoses of necrotising spider bites started spreading around New Zealand too. There was one common factor about all of these misdiagnosed spider bites – they didn't match the three criteria for a conclusive diagnosis. The bites weren't witnessed, the spiders weren't collected, and the spiders weren't identified by a professional arachnologist.

The case was definitively closed on this myth in 2003 when two scientists (clinical toxicologist Geoffrey Isbister and arachnologist Michael Gray) investigated every definite white-tailed spider bite in Australia over a three-year period. They tracked every reported spider bite in Australia and found every case where the bite could be unquestionably attributed to a white-tailed spider. Guess how many of these white-tailed spider bites resulted in necrotic ulcers – absolutely none. The symptoms of the bites were the expected local pain and swelling that went away after a day or so. There have been a number of other studies into the chemistry of the white-tailed spider's venom that haven't found any evidence of a substance that would cause necrosis. You'll remember from the last chapter that there are a few spiders in the Sicariidae family that are known to cause necrosis due to the presence of phospholipase D compounds in their venom. Neither this enzyme, nor any other necrotic enzyme or bacteria, has ever been found in white-tailed spider venom.

The myth of necrotic white-tailed spiders seems to have stemmed from a small number of imaginative doctors jumping to conclusions. When presented with a mysterious case of skin necrosis, some doctors seem to have landed on spider bites as a 'putative' or 'probable' cause in the absence of any other definitive diagnosis. We can only guess as to what led to these initial misdiagnoses. Perhaps the doctors had heard of recluse spiders and, not being trained arachnologists, assumed that this characteristic could be easily applied to a completely different spider on the other side of the planet. Perhaps their own spider fears, or even the fears of the patient, swayed their judgement. In some cases, patients mentioned having seen white-tailed spiders in the house, and that seems to have influenced the misdiagnosis. I hardly need to point out how spurious this connection is. White-tailed spiders are such common household visitors that you may as well conclude that coffee cups cause necrosis because the patient saw one in the cupboard this morning. Even the published medical papers erroneously diagnosing these spider bites point out that the bites weren't witnessed and in most cases the patient didn't even recall ever feeling a bite happen. Worryingly, it has been suggested that administrative processes surrounding medical insurance can lead to misdiagnosing spider bites, as there is pressure on doctors to put a diagnosis down on paper, even when the evidence is not there.

Despite the myth being clearly debunked years ago, something about it is tenacious and it continues to persist. In 2017, over a decade after the myth-busting work of Isbister and Gray, a story hit the newspapers about a man in Victoria with a serious necrotic infection that required both legs to be amputated. According to the man's family, the doctors had told them it was likely due to a spider bite despite the man not recalling ever being bitten by a spider. From here on in the media did what the media does and the myth of the necrotic white-tailed spider reared its ugly head again.

Clearly, this is a case of spiders' bad reputation taking

precedence in people's minds over evidence. And it's not just white-tailed spiders that have shouldered the blame for necrosis. *Badumna* spiders (small black house spiders) have also been falsely accused of causing skin necroses. In the 1920s a wave of skin necroses was erroneously blamed on wolf spiders. Yellow sac spiders (*Cheiracanthium*) and hobo spiders (*Eratigena agrestis*) have all been falsely accused of causing the same condition.

The consequences of misdiagnoses are far worse than a few bad newspaper articles. Skin necroses can be a symptom of a long list of other conditions including bacterial, fungal and viral infections; circulatory and inflammatory, and sometimes more serious underlying conditions like diabetes; and certain cancers. In a survey of supposed necrotic spider bites in New Zealand, testing found that most of the ulcers were *Streptococcus* and *Staphylococcus* bacterial infections. In 1990 the *Medical Journal of Australia* published a report on a man who was erroneously diagnosed with 'necrotising arachnidism' and prescribed painkillers, antibiotics and anti-inflammatory drugs. After weeks of this treatment the patient's symptoms weren't improving. Eventually, after more testing, the actual cause was found, and the man's necrosis was found to be a symptom of a rapid-acting leukaemia that affects the blood cells and bone marrow. Where potentially serious conditions are misdiagnosed as 'necrotic arachnidism' it can prevent those patients from receiving the appropriate care for the condition they actually do have. Even in areas where necrotic spider bites may occur, such as in the southern United States where encounters with brown recluse spiders are common, misdiagnoses of necroses are common. Spider bites are rare, even from brown recluses, and only a small proportion of them lead to severe envenomation and necrosis. Doctors are advised to investigate more likely causes where a conclusive diagnosis of spider bite can't be made.

Daddy-long-legs spiders *are not* the most venomous spiders in the world

Let's make sure we are clear about what we mean by 'daddy-long-legs' spiders. In parts of Europe, daddy-long-legs is used to refer to long-legged craneflies. They look a bit like mosquitos on steroids. In North America daddy-long-legs may refer to small animals called harvestmen. These are not spiders, and are not venomous, though you couldn't blame anyone for mistaking them for spiders: they have eight legs and are closely related to spiders. They belong to a group called Opiliones and, along with things like scorpions, ticks and mites, are another type of arachnid. More often than not the common name daddy-long-legs refers to a group of spiders also known as cellar spiders, carpenter spiders or by their family name Pholcidae.

You have probably encountered daddy-long-legs spiders in your house. They are the most frail looking, harmless spiders imaginable, so it comes as a surprise when you hear the myth that they are the most venomous spiders in the world. This is often told with the caveat that their fangs are too small to be able to puncture human skin. The myth has such a tantalising idea behind it: that the deadliest venom on the planet is hidden inside fragile creatures that wander around our homes and would kill us all if it weren't for their meek little fangs. It teases us with the skin-crawling possibility that one day, one boisterous daddy-long-legs could finally break through our tough leathery skin, just a little bit, unleashing untold havoc.

And that spider could be the one under your couch right now. It sounds fun, but it's bollocks.

It's unclear where this myth has come from but one possibility that has been considered is that it was inspired by the daddy-long-legs' predatory behaviour, which was described in the last chapter. Their webs are very good at catching other spiders, even species venomous to humans like *Latrodectus* red-back spiders. Perhaps observing a red-back being taken down by a daddy-long-legs led someone to imagine that anything attacking a venomous red-back would need even more potent venom still. As we covered in the last chapter, this isn't how venom works. It's likely that the complex webs of pholcid spiders are doing most of the grunt work in incapacitating other spiders, and their venom is only needed to put the finishing touches on a successful prey capture. Recent studies have shown that daddy-long-legs venom is actually quite weak, even against small insects.

It should come as no surprise that this idea is false, especially since there was a great big spoiler in the last chapter where we talked about Sydney funnel-webs being the most venomous spider in the world, and the issues with defining what 'most venomous' actually means. It's an idea that has been debunked many times – there was even an episode of *MythBusters* about it, where the hosts volunteered to be bitten by daddy-long-legs spiders, which clearly disproves the other half of the myth: that their fangs can't penetrate human skin. Typical symptoms of a daddy-long-legs spider bite in humans are a mild and short-lived sting.

Before we close the lid on this myth, I have one final thought: the more I say 'daddy-long-legs', the creepier it sounds.

Like, whoever named them had a weird fetish thing going on. Is it just me, or does including the word 'daddy' seem a little strange and unnecessary. Just me? OK, moving on.

Spiders *do not* live in your hair

This myth has been around a long time and is definitely an urban myth. It's the kind of tale that usually gets told as a spooky story around a campfire, but unfortunately is re-told every now and again as if there is a sliver of truth to it. The version I remember hearing as a kid was about a woman who had a huge 60s-style 'beehive' hairdo, held together with copious amounts of hairspray. To preserve her glamorous hairdo, she never washed it or combed it. At some point, a spider found its way inside the hair and began laying eggs. One day the woman suddenly dropped dead, and when the doctors cut open her hairdo (which I guess is something you might do in an autopsy?) they found hundreds of baby spiders living inside her hair and biting her scalp.

There are many versions of this tale, some involving other elaborate hairstyles like afros or braids. When I was in my early twenties I, like so many other super-cool dudes, went through the dreadlock phase. I must have heard similar apocryphal stories daily about people with spider nests in their dreadlocks. This one is pretty simple to debunk: there is nothing to it. No medical records, no newspaper reports, no misinterpreted spider behaviours that would lead people to think this. It's just a fun creepy story that is completely untrue.

The email/meme/post you saw about deadly spider attacks is *not* true

I'm old enough to remember a world without the internet. When it spread across the world and found its way into our homes, it came with the promise of instantly sharing information across the globe. At the time, I think many of us assumed that meant good, factual, robust information. Our assumptions were very quickly proven wrong. Myths continue to spread rapidly throughout the internet, and our unfortunate spider friends have been at the centre of many of them. I remember the days of chain-emails telling some long story from a 'friend of a friend', ending with explicit instructions to forward this email to everyone you know for their own good. Later came the viral memes with pictures of some random spider, or something that looked vaguely like a spider, accompanied by some made-up 'fact' about how deadly and voracious it is. Now in the age of fake news and social media it's a safe bet to assume that anything you read on the internet about spiders is potentially untrue.

It's hard to point to one single spider myth that has taken the internet by storm, because there are so many. One that comes to mind was a story circulating on social media about camel spiders that were terrorising US troops stationed in Iraq. They were said to be highly venomous and flesh-eating, and were called camel spiders because they laid their eggs under the skin of live camels. The story was always accompanied by an image of camel spiders that either used photo-editing or forced perspective to make them look the size of a small dog. Camel spiders, despite their common name, are not even spiders. Like harvestmen, they are another group of 'non-spider' arachnids (order Solifugae) that have no venom whatsoever, and do not do any of the horrible-sounding things they are accused of. They can grow quite large, but only up to around 15 centimetres in length, not the size of a chunky Pomeranian as suggested by the myth. Whenever I see one of these fake articles, or any widely circulated rumour on the internet, I always ask the same question: Where did it come from? Surely someone had to be the first link in that chain-mail or meme, who made it their business to photoshop together some blurry images and make up a fake story about spiders to send out into the world. But why? What do they get out of it?

In 1999 the Department of Entomology at the University of California suddenly started getting lots of public enquiries concerning a story that was circulating about the deadly South American blush spider. The source of this public concern was an email hoax that was being frantically sent around by friends and families concerned for each other's safety. The email described, in great detail, recent deaths supposedly caused by the blush spider. It went into all manner of specifics about an article published in the *Journal of the United Medical Association* that described the species *Arachnius gluteus*, and its toxic venom and bite symptoms. The email talked about how the Civilian Aeronautics Board was grounding flights after spider nests were discovered under plane toilet seats, and how several people had died after being bitten in the toilets at Big Chappies restaurant at Blaire Airport in Chicago. It didn't take much effort to recognise a hoax. There is no such thing as a blush spider, not even a genus called *Arachnius*, there is no *Journal of the United Medical Association*, no restaurant called Big Chappies anywhere, especially not in Blaire Airport, which also doesn't exist, and there's no such thing as the Civilian

Aeronautics Board. Nevertheless, the email continued to spread. Scientists at the University of California took it upon themselves to make a website dedicated to debunking this specific email hoax, where they could redirect all the public enquiries they were receiving. The debunking website itself went viral and received almost 50 000 visits within two weeks of going live.

The question still remained – why? What maniac would have bothered to go to all that trouble to write such a detailed story with so many falsehoods in it? Well, a week after the debunking website went live, the scientists received an email from that very maniac, who turned out to be just a very reasonable and curious person. He was simply a regular guy who was asking the same question: Where does this false information come from and why does it spread so rapidly through the internet? He decided to test whether he could start an internet hoax, and boy did he succeed. Assuming that people would be drawn towards a scary spider story, he penned the tale of an exotic-sounding spider that crawls out of toilets and bites people on the buttocks (hence the species name '*gluteus*') and sent it off to about 30 friends and family members, complete with as many utter lies as he could fit into the story. He suspected that people were incredibly gullible, and sadly he was overwhelmingly proven right.

Why do spider myths persist?

The internet has changed a lot since the days of chain-mail hoaxes, but its ability to spread misinformation has not. It has arguably gotten worse with the rise of social media and our ability to be constantly interacting with misinformation through smartphones and other ubiquitous devices. There is a well-known psychological phenomenon called 'negativity bias', where we are more likely to focus on information that is phrased negatively or that uses negative emotional language. This is nowhere more apparent than on social media, where anything that scares, annoys, angers or outrages us is more likely to be engaged with and shared. Studies have shown that even when a news story is positive, articles that take a negative spin, or use negative language, are more likely to be shared online than those using positive language. This is exacerbated by social media and website algorithms that aim to increase user engagement and time spent on the platform by showing you content that you are more likely to engage with. Inevitably this means putting more and more negative and outrage-inducing content in front of you. Negativity bias is not a new phenomenon, nor is it unique to the internet. The old journalist saying 'if it bleeds it leads' shows that journalists have always known that outrage and negativity are the secret to writing a popular news article. This doesn't provide much incentive for journalists and media outlets to prioritise accuracy over sensationalism. Combine negativity bias with people's emotional responses to spiders and it's no surprise that the internet can turn into a dumpster fire of spider misinformation.

Recently, a global team of scientists showed that the less objective and accurate spider news articles are, the more likely they are to be shared online. They compiled an enormous database of every spider-focused news article published online over a three-year period (over 5000 articles from 80 countries in 40 languages) and assessed every article based on its accuracy and how sensationalist the language was. Disappointingly, almost half of the articles were found to have factual errors, and the articles that used more sensational language tended to have more errors. As would be expected, articles using emotive and overblown language were more likely to be engaged with and shared. One positive finding was that articles that contained input from spider experts were less sensational and more accurate. Sadly, the same wasn't true for articles with input from medical and pest-control professionals. The researchers actually pointed the finger at medical professionals for contributing to the spread of misinformation about spiders by providing inaccurate identifications of spider species, and misdiagnosing spider bites. These same patterns were found in a separate study investigating every piece of spider news published in Italy over an entire decade. In this study, a whopping 70 per cent of articles contained factual errors, including articles that gave the wrong species name or even showed photos of the wrong spider species. Again, articles that were alarmist and sensational were more likely to be shared online than those that were more objective. Sensationalism was ranked based on the use of emotive words in the article, and some of the more ludicrous examples the contributors to this study listed are articles with the words 'devil', 'nightmare', 'agony' and 'terror'. They noted that virtually none of the hundreds of articles published that decade had input from spider specialists, and that none of the spider bites reported were verified by the spider being collected and correctly identified. So, based on our now very familiar spider-bite diagnosis criteria, there wasn't a single verifiable spider bite in Italy that entire decade. Furthermore, the articles covered three supposed deaths from spider bites. Two of these were later shown to be fake news stories and the third couldn't be verified with follow-up investigations. To summarise, spiders and the media make embarrassingly bad companions.

Given the ease with which misinformation spreads, and many people's emotional responses to spiders, you can't really blame people for falling for the odd spider myth here and there. But we also can't blame the spiders: they are the innocent victims being unfairly portrayed because of our subconscious tendencies. The onus falls on us to tell better spider stories and to resist repeating the same mistakes of sharing alarmist spider headlines and outrage-inducing fake news. So, let's get back to telling better spider stories. Who needs fake news when many of the real facts about spiders are arguably more far-fetched and mind-boggling than the myths? Of all the hard-to-believe facts you will hear, many of them have to do with spider silk. You might have read some of the hyperbolic stories about tougher-than-steel spider silk floating around the internet. It turns out that these are actually true. Silk seems like such a simple substance but is so sophisticated and complex that it is, almost, unbelievable.

Chapter 6

EIGHT-LEGGED BIOENGINEERS: THE SCIENCE OF SILK

One of my earliest experiences studying spiders was when I was a student volunteer in a lab hidden in a university basement. The spider lab was situated in between a human blood protein lab and a cobalt radiation storage facility. If I was ever going to get superpowers from an unfortunate lab accident, this was where it was going to happen. Sadly, no matter how clumsy a lab assistant I was, I wasn't bitten by a single radioactive spider. Perhaps the most disastrous accident that happened in the lab was when a few lids were left loose on the fruit fly enclosures. For days afterwards the basement lab was filled with clouds of small annoying flies who had escaped their fate of becoming spider food.

It was my job to clean out the orb-web spider frames. They were simple enclosures made from a square plastic frame, about 30 centimetres across. The inside was lined with rough masking tape where spiders could attach webs. Once you put a spider in the enclosure it would pretty soon build a web that sat nicely in the centre of the square frame. The frames didn't have a front or back; they slid neatly into a little chamber on a shelf so that when you wanted to observe each spider you just pulled out its frame and the web would come with it, with the spider sitting still in the centre. I would carefully sweep around the web to remove any fruit fly remains, and give the spider some water with a gentle misting from a spray bottle. After the escape of the fruit flies, this job also involved trying to keep calm and not go mad while a never-ending plague of fruit flies crawled over your face, in your eyes and up your nostrils.

One morning after the fruit fly escape, I walked down into the lab and heard a long string of unrepeatable profanities being screamed at the incessant fruit flies. I poked my head into the room where we kept the spiders to find one of the visiting researchers trying to get work done while also slowly going insane. In an attempt to combat the fly swarms he had built himself a defensive wall of spider webs. He had taken out each of the plastic frames and stacked them all in a giant grid reaching all the way from his desk to the roof. I hadn't met this person before, so we shared an awkward greeting through a wall of spider webs before I left him to finish whatever he was doing inside his silk bunker. I don't think his plan of intercepting the fruit flies with orb-webs was particularly successful as the swearing continued for at least a few hours. Regardless, I thought it was an ingenious use of spider silk, and given our track record of trying to find practical uses for spider silk this could be one of the more successful stories.

To some, spider webs are simply seen as a garden nuisance, or a small blemish in a well-kept home. To others, spider silk represents the holy grail of bioengineering for its unmatchable material properties. To spiders, silk is everything. Silk is a sensory tool used to send and receive information. It builds shelters to live in, and traps to hunt with. Silk entangles writhing prey, and caresses vulnerable eggs. In a way, spiders and their silk are indivisible. Think about the familiar orb-web spider, who uses her web to listen to the world around her. Each morning she builds a web using silk made from glands in her abdomen. At the end of the day she takes that silk back in, consuming the precious nutrients so that they can be used to build tomorrow's web. Day after day her web acts like another organ that expands and contracts in a daily rhythm. Since spiders use their webs to listen, it has been argued that silk is an extension of their central nervous system, linking seamlessly with their sensory organs and nerve cells.

And there isn't just one type of silk. Different species of spiders use different types of silk, and individual spiders can spin different types of silk depending on what that silk is to be used for. In the abdomen of each spider is a miniature silk factory that can conjure a cocktail of different silk types on demand. The complexities of this chemical and biological process are only just beginning to be understood, and they are far more intricate than anything humans are able to replicate. That hasn't stopped people from trying, and our attempts at harnessing the power of silk are equal parts exciting, frustrating and downright ludicrous.

What is silk?

Spider silk is constructed of proteins made in glands inside the spider's abdomen. Each spider has up to seven different glands that each produce a distinct type of silk protein, which have all been given fancy names like 'major ampullate silk', 'minor ampullate silk', 'flagelliform silk' and 'cribellate silk'. These glands provide all the ingredients spiders need to spin many different silks with wildly varying properties. For example, if an orb-web spider was to make some tough dragline silk to lay down the strong foundations of a web, it would mix together some major ampullate silk proteins from the major ampullate gland, with a few pinches of minor ampullate silk proteins from the minor ampullate gland. If that same spider then made silk for the spiral threads of their web, they would take some flagelliform silk protein, coat it with some aggregate silk proteins, and sprinkle it generously with glue droplets, to make some stretchier, stickier silk perfect for catching prey.

The silk from different glands is actually mixed together outside of the spider's body. If you were to use a high-powered microscope to watch silk being reeled from the spider's body, you would see that each silk gland has a separate opening and each separate silk type is pulled out of the body in multiple threads, like noodles of fresh pasta being pushed through a spaghetti machine. As silk exits the spider's body it goes through a chemical change from liquid to solid. The pressure applied to the fluid as it passes out of the glands through microscopic channels actually shapes and aligns the protein molecules. Once outside, each silk type begins to interlock and crystallise, forming a solid thread. The temptation to liken a spider weaving silk to a weaver spinning threads may actually be quite apt, as there seems to be some element of craft to it. The threads of silk are guided out of the spider's abdomen by small pairs of legs called spinnerets. Spiders will also use their walking legs to pull the silk thread from their abdomen. By manipulating the speed at which the silk is reeled, spiders can modify the strength of the web. Orb-web spiders that reel their silk quickly create stronger silk than when it is reeled more slowly. Using a small number of ingredients, combined with some complex nanoscale chemical processes, silk is to a spider what a knife is to a Swiss soldier ... You know, a Swiss Army knife. Silk is the ultimate tool; it can be built on demand and adapted to any number of purposes.

Of all the different types of silk that spiders make, the dragline silks (also called major ampullate silks) spun by orb-web spiders have been studied the most intensively. These are the strong, taut silks that form the radial threads of orb webs, and are also used as safety lines by spiders as they climb, to provide a safeguard against falling. In a web, dragline silks absorb the impact of flying prey, so they have evolved to have maximal strength and elasticity. For this reason, they are the most likely candidates for holding secrets that could help develop super-strong, bio-inspired materials. Silk is an engineer's dream; it's incredibly strong while still being lightweight and flexible. Weight-for-weight, spider silk can be stronger than steel and can absorb more energy than a bulletproof vest. There is one spider whose silk stands out above them all; its dragline silk is stronger than any other material in the natural world.

The strongest biological substance in the world

In the early 2000s a team of scientists set out on a mission to find the world's toughest spider silk. This is no simple task considering there are over 50 000 species of spiders worldwide, all of which use silk in some capacity. Rather than going through every different type of spider one by one, trying to find the toughest silk on the planet, the scientists narrowed their search using what they already knew about spider behaviour. They ruled out the spiders that don't use silk for catching prey. The fluffy silk that spiders use to carry around their eggs, for example, probably isn't going to have the most hard-core strength. Instead, they went straight to orb-web dragline silk, which narrowed their search down to around 3000 species. From here, they reasoned that whichever orb-web spider has the toughest silk, it would probably have the biggest web, which would be used for taking down some pretty big prey. They scoured the literature, searching books and expedition records for stories about enormous spider webs. And in this search, one spider stood out among them all.

Darwin's bark spider (*Caerostris darwini*) is found in the jungles of Madagascar, where they don't just spin webs in the spaces between trees, they spin webs across entire rivers. When the team of scientists headed to Madagascar in search of Darwin's bark spiders, they measured webs spanning rivers up to 25 metres across. To put that into perspective, that's longer than a semi-trailer, longer than a cricket pitch, or five times as long as a giraffe is tall. In other words, it's bloody huge. Strung taut across rivers, these webs are probably catching large flying insects, like dragonflies, beetles and moths. At this size I wouldn't be too surprised if they caught the odd fisherman here and there. To measure the strength of this spider's silk, the team took samples of the thick radial strands that scaffolded the bark spiders' webs and tested it using the same kind of machines used to test the structural integrity of building materials. The numbers that the team measured were off the charts; Darwin's bark spider silk was twice as tough as any other spider silk – it absorbed ten times more kinetic energy than bulletproof Kevlar. They hadn't just discovered the toughest spider silk; they had found the toughest material produced by any living creature on the planet.

What makes spider silk so fascinating is not just that it's strong, but that it is elastic and flexible at the same time. It's also self-assembling, recyclable and biodegradable. It's these combined qualities that make engineers drool when they see a spider's web. Yes, concrete and steel are incredibly strong, but can you bend and stretch them into different shapes? Yes, the carbon fibre chassis of your car can absorb massive amounts of energy, but does your car bounce back into shape after a crash? Spider silk does all of this and more. And to think it just oozes out of the backsides of little crawly things in our garden.

Keep in mind, when we talk about the incredible properties of spider silk, we are talking in relative amounts. So, the amazing strength of a thread of silk a few hundred nanometres thick, is only amazing in comparison to a thread of steel a few hundred nanometres thick. For us to fully harness the potential of silk, we would need a way to generate it in large industrial quantities. Now, I know what you're thinking. Farming spiders? Ha! What a silly idea. No-one would ever be crazy enough to try to milk silk out of spiders. It's ludicrous, right? Right!?

Weaving spider silk

Jacob Paul Camboué was a French missionary who travelled to Madagascar in the late 1800s. At some point during Father Camboué's mission he encountered local golden orb-web spiders and, whether it was divine inspiration, or perhaps just idle hands, decided to embark on a new mission: to prototype a new textile manufacturing process using harvested spider silk.

Golden orb-web spiders (genera *Nephila* and *Trichonephila*) are builders of enormous webs. They are often glossy black with splatters of yellow, white and red all over their bodies. Their common name comes from their thick silk, which glows bright gold in the sunlight. Some species can grow about as big as your hand and their webs can be several metres across. Their webs are large enough, and their silk strong enough, that the odd bird, frog or snake gets tanged in one and becomes a spider's lunch. If anyone was ever going to decide to make textiles from spider silk, the tough and shimmering gold silk of *Nephila* spiders was a good place to start.

Camboué started experimenting with silk by collecting large webs and trying to spin threads from the tangled masses. The results were pretty messy and so he changed tactics to see if he could draw silk directly from the spiders. He started by taking a single spider and enclosing its head and legs inside a match box. With its fangs and legs out of the way, Camboué would give the spider's abdomen a small squeeze to make it release a small thread of silk. By pulling gently on the thread Camboué could reel out long single strands of silk. He refined this technique by taking a small hand-cranked reel that he would attach the silk to and slowly wind it up, like thread on a bobbin. From one spider, he could get single strands of silk hundreds of metres long.

With a successful prototype in place, it was time to start scaling up Camboué's marvellous spider-milking machine. He worked with a professional school in Madagascar's capital city Antananarivo. Madagascar had only recently been taken over and established as a French colony, so the professional school was built with the very colonial mission of teaching the locals how to be a little more French and a little less Malagasy. With France being a global hub for silk manufacture, it made perfect sense to make this new colony, and the professional school, the launching pad for France's next silk revolution. The new field of 'araeniculture' and the fabled luxury of spider silk products were poised to take the world by storm. And this was all going to be made possible by the invention of a silk-harvesting machine known by the very ominous, and very French, name: 'the guillotine'.

The guillotine could harvest silk from 24 spiders simultaneously, thus spinning threads of silk 24 strands thick. Each spider was tethered in a small box. A strap ran across the spider's head and legs, stopping it from interrupting the flow of silk. The similarity to a person's head being held in place at the base of a guillotine is obvious and I imagine this is where the machine got its name. The 24 spider boxes were arranged in a vertically stacked grid, and each of the 24 silk strands was drawn through a central ring and attached to the hand-cranked reel. Malagasy girls attending the school were sent out into nearby parks with baskets to collect spiders by the hundreds. A single spider could be used to harvest almost two kilometres of silk before needing to be released back into the wild. After a few weeks, the same spiders could be collected again for harvesting. For months on end the Malagasy girls were busy harvesting and spinning the raw spider silk. And after that the girls were put to work at looms, weaving the silk threads into fabric. After all this effort it was estimated that the factory could produce just 365 metres of silk thread a month.

Finally, in 1900, the work of Father Camboué and the professional school hit the world stage. At the 1900 Paris

Illustration of the 'guillotine' used to harvest silk from golden orb-web spiders in Madagascar in the late 1800s
'The silk-producing spider of Madagascar' (1900). *Scientific American 83* (9), 133.

Exhibition – a world's fair showcasing the latest artistic, scientific and cultural developments – Madagascar displayed a complete set of luxurious bed-hangings made entirely from golden spider silk. News spread across the globe of the remarkable guillotine invention and the dawn of a new textile's era. Camboué began shipping Malagasy spiders back to France so that silk production could continue in the city of Lyons, the European centre of silk manufacture. There were even reports that Camboué and the professional school were working on an experimental covering for a hot air balloon to be used by the French Military Balloon School in Calais.

Camboué wasn't the only one who had tried to create textiles from spider silk. The French have had a particular interest in such textiles for a long time. Almost two centuries earlier, Monsieur Bons, the President of the Royal Society of Sciences in France, wrote a paper on the 'usefulness of the silk of spiders', arguing that spider silk could be just as useful as that of silkworms, from which silk textiles were traditionally made. Silkworms aren't actually worms; they are the larvae of the *Bombyx* moth, which weaves a thick, fluffy silk cocoon. Monsieur Bons reasoned that since silk was harvested from silkworm cocoons, he could also harvest the fluffy silk of spider egg cases that looked somewhat similar. He managed to collect about 85 grams of spider egg cases, which he cleaned and spun into a silk 'much stronger and finer than ... common silk'; it was used to make a pair of gloves and a pair of stockings. Around the same time, French naturalist René Antoine de Réaumur was tasked by the French government to investigate practical uses for spider silk. He built upon the work of Monsieur Bons but eventually concluded that spider silk used in this fashion would not be commercially viable.

Attempts to harvest silk directly from spiders' abdomens didn't come until much later and, surprisingly, there have been multiple independent inventions of silk-harvesting machines. Ramón M. Termeyer, a Spanish naturalist, seems to have taken great offence to Réaumur's dismissal of spider silk as a commercially viable product. In the 1770s Termeyer built a small device much like Camboué's, where the spider's head and legs were restricted and silk was drawn from the abdomen by a slowly spinning reel. In confidently rebutting Réaumur's claims, Termeyer confidently concluded that 'greater profit can be drawn from spiders than from silk-worms'. In the 1820s, Daniel Brandson Rolt, a factory worker from London, discovered that he could draw a thread of silk from a spider and attach it to a small steam engine that would reel the silk, thus inventing the first steam-powered silk harvesting machine. Using common garden spiders, he could use the machine to reel 45 metres of silk per minute. These inventions were unknown to Burt G Wilder, a US Army surgeon and Civil War soldier, who in the 1860s developed and patented a hand-wound silk machine strikingly similar to Termeyer's that he used to collect the silk of *Nephila* spiders.

After centuries of excitement about the possibility of a new spider silk textile industry, why aren't we all wearing spider silk gowns, and what happened to the great araeniculture revolution? There are a few simple hurdles that have stopped spider silk from becoming a viable industry. Rearing spiders in mass quantities is pretty hard work. We can't exactly put spiders out to pasture in big herds without them eating each other, so they have to be housed and fed individually. Then they have to be fed live prey, which means we have to provide them with enough space to build a prey capture web. Breeding spiders is an even trickier business, what with the whole sexual cannibalism, infanticide and siblicide thing. Finally, the logistical effort required to harvest spider silk is enormous. It's really fiddly and, more importantly, there's just something creepy about spider stirrups. Contrast this with silkworms, which will quite happily live together in the thousands, engorging themselves on mulberry leaves until they are big enough to build a thick, soft silken cocoon. All things considered, silkworms are simply more cost and labour effective. When Camboué was developing spider silk textiles in Madagascar, it was estimated that just one pound (453 grams) of spider silk was worth £8, which in today's equivalent is somewhere around €700, or A$1100. Even Burt G Wilder, after years of developing his silk harvesting machine, eventually concluded that these inventions were 'what sober, cautious men already expect of it – a means of luxury, of comfort and of national wealth'. Despite the valiant efforts of a few, the spider silk industry was short lived. There are no surviving examples of Father Camboué's spider silk textiles, and one can only imagine what the bizarre course of history would have been, had we lived through an era of French spider silk war balloons.

Thanks to the work of entrepreneur Nicholas Godley and textile designer Simon Peers, we don't have to rely on our imaginations to know what spider silk garments look like. In 2004, inspired by the tales of father Camboué, Godley and Peers teamed up to recreate the fabled guillotine and see if they could create their own spider silk textiles. They headed off to Madagascar, tracing the footsteps of Camboué and, using records and descriptions of his original machine, began working on a new and improved silk harvesting machine. Peers had spent decades working with local Malagasy textile manufacturers, and assembled a team to begin harvesting and spinning silk just like the Malagasy weavers did over a century earlier. In 2009, Godley and Peers announced that the weavers had produced the only spider silk textile of its kind in the world, a four-metre-long *lamba akotifahana* – a traditional Malagasy shawl. It went on display at the American Museum of Natural History in New York, and just as the legends had told, the silk's natural colour was a bright shimmering gold. Three years later, the team of weavers had produced their second garment – a golden cape, intricately embroidered with images of spiders and plants. It went on display at the Victoria and Albert Museum in London and has toured the world ever since. Those lucky enough to have touched, or even worn these rare fabrics, have remarked on how they feel unlike any other textile – impossibly soft, almost weightless, but incredibly strong – exactly what we would expect from a garment made of spider silk. It's estimated that this cape used the silk of over 1.2 million spiders, and the work of over 80 weavers and spider collectors.

In addition to the prohibitive cost of making spider silk textiles, the silk exhibits something called 'supercontraction', where silk threads contract (a lot!) when they get wet. So, hypothetically, if someone wearing a spider silk shirt got caught out in the rain, their stronger-than-steel-spider-blouse could soon turn into a shimmering gold boa-constrictor, crushing their ribs and suffocating them. So, yeah, that's a bit of a problem. After centuries of attempts to farm spider silk, it still remains elusive and impractical.

The next logical approach to harnessing the power of silk was to find a way of recreating it artificially. If we could recreate the structural properties of spider silk with an artificial material, and manufacture it in large quantities, we could create these fabled super-strong, ultra-lightweight materials, and save millions of spiders a lot of discomfort. It should be pretty simple, right? After all, it's just goop from a spider's butt. In reality it's much more complicated, and the ways that we have tried to do this can seem pretty strange. If you thought stories about Jesuit priest spider-farmers were weird, it's about to get weirder.

Decoding spider silk genes

In 1990, scientists sequenced the first spider silk protein gene. Again, the focus was on the super-strong dragline silk of *Trichonephila clavipes*. They used the tried and tested method of reeling dragline silk from live spider abdomens with a small electric motor, and collected cells from spider silk glands to figure out what amino acids formed the building-blocks of this protein *and* the genetic code that contains the instructions for building it. Of course, as with almost everything in biology, there is never just one magic silk gene, especially not for something as complex as spider silk, but it was the first step towards solving one of the peskiest problems with harvesting spider silk – the spiders. Since spiders don't give up their precious silk without a fight, the plan was to use rapidly advancing genetic technologies that would allow us to read the blueprints written in spider DNA and create artificial silk in a laboratory. Spider silk proteins were given the fancy name of 'spidroins', and once we had the first spidroin gene sequence, research advanced pretty quickly. After a few years we had spidroin gene sequences for different types of spider silks and from different spider species.

Scientists are now beginning to understand the chemistry and physics of how spider silk works. For example, the proteins (or spidroins) that are the main ingredient of dragline silk are made of repeated chemical structures that stack on top of each other, end to end, like little molecular building blocks, or links in a chain. The chemical bonds between these building blocks are very difficult to break and so it takes a lot of energy to snap the chain of molecules, which is why dragline silk is so strong. This is why spider silk is described as 'crystalline' – the molecules arrange themselves together in a very ordered pattern, like the precisely arranged minerals that form crystals. Other types of spider silk are made up of different types of proteins that have different properties. Flagelliform silk, which forms the stretchy spirals of a spiders' web, is made up of protein that has a different amino acid structure. Rather than stacking together, these chemical structures form spirals, which explains why flagelliform silk is so elastic – it's literally made out of nano-springs, that can stretch out and pull themselves back together into their original configuration. This kind of information is what gets materials scientists excited, thinking about the bioengineering possibilities of silk. The more silk types that are studied, and the more protein types we understand, the closer we get to the dream of putting together some form of 'spider silk cookbook', where proteins with various properties could be mixed together to build all manner of crazy biomaterials. But deciphering the genetic and chemical recipes for these proteins is the easy part. Things get more complicated when it comes to building them from scratch in a laboratory.

Brewing artificial spider silk

Within a few years of the first spider silk genes being discovered, scientists made the next big leap in artificial silk manufacturing by taking those genes and inserting them into *Escherichia coli* – the bacteria you have probably heard referred to as simply *E. coli*. While they are famous around the world for making themselves at home on dirty kitchen benches and uncooked meats, *E. coli* bacteria are also famous for their usefulness in genetic experiments, particularly as hosts for other creatures' genes. These bacteria are easy to grow in the lab and have a small and simple genome, so these days it's rather easy to take the genes from another organism and insert them into the genome of *E. coli*. Relatively speaking of course – it's easy when done by highly skilled geneticists in a well-equipped laboratory, not by you in the kitchen sink. With a few modifications to the spider silk genes to make them more compatible with *E. coli*, they were inserted into the bacteria, and the bacteria were left to grow and multiply. As the genetically modified bacteria were swimming around in their bacterial soup, doing whatever it is bacteria do, they also started oozing out spider silk proteins. Initially the amounts of silk they made were low, and the sizes of the proteins were much smaller than what you would expect from real spider silk. Since then, synthetic biologists have been refining the laboratory protocols and making edits to the spliced gene sequence to improve the yield and quality of artificial spider silk.

Compared to harvesting silk directly from live spiders, growing transgenic silk in *E. coli* is much easier to scale. The process isn't too dissimilar to brewing beer. If you have ever dabbled in home brewing you would know that the fermentation process is pretty simple: you get some yeast cells, put them into a giant vat of water, keep them warm and give them plenty of food (sugar) so that they grow, multiply, and spit out alcohol as a by-product of their metabolism. Similarly, to brew spider silk, you take these *E. coli*, feed them in a big warm vat, and while they are growing and multiplying, they also make silk proteins based on the spider genes spliced into their genome. Let's be clear about something: from here on in, we're not really talking about spider silk any more. We're talking about transgenic silk, which oozes out of fermenting bacteria that have been genetically modified with an artificially created gene. Which is still pretty cool. The ability to brew these proteins in large vats has opened up a wide array of new possibilities and spider silk has been touted as the key to new bio-inspired textiles, bulletproof vests, parachute cords, fishing lines, lightweight aviation parts and many other tantalising possibilities.

It occurs to me that there is one industry that isn't making use of this technology but probably should – the craft beer industry. Before spider silk genes were put into *E. coli* bacteria, scientists successfully made artificial silk by inserting spider genes into yeasts. Admittedly these weren't brewers' yeasts, but the principle is there. Now, I may not be a synthetic biologist or a brewer, but as far as this armchair observer can tell, there's a bright future ahead for the spider silk craft beer industry. Any keen start-ups interested in disrupting this industry are welcome to contact me for market research and product testing.

One downside to making artificial silk using vats of *E. coli* is that, unlike real spider silk, it doesn't come out of the bacteria in nicely formed threads ready for weaving. Just as the spider uses its spinnerets and gland ducts to physically manipulate silk proteins into the right form, we've developed our own processes for spinning artificial silk. In spiders, the silk proteins start off inside silk glands and are pushed through a network of ducts and spigots before forming solid threads when released from the spider's abdomen. Artificial silk proteins produced by *E. coli* start off floating around in a solution of bacterial broth, so we need a rather different process to spin them into fibres.

First, the silk proteins have to be isolated from the bacterial broth and with some chemical wizardry they can be kept dissolved in a concentrated fluid solution called spinning dope. After this step a number of different spinning methods can be used to turn the dissolved proteins into silk fibres. Techniques such as wet spinning, dry spinning, and electrospinning all rely on different ways to rapidly evaporate the fluid that the proteins are dissolved in. With the fluid removed, the silk proteins will rapidly self-assemble into thin strands. What technique is used depends on the type of spinning dope you begin with, and what material properties you want the artificial silk to have. Different spinning methods can have just as significant an impact on the thread quality as the silk proteins they're made of.

Even though *E. coli* is arguably the most common source of artificial silk, it isn't the only organism that's been experimented with. Scientists have tried splicing spider silk genes into yeasts; *Salmonella* bacteria; and potato, tobacco and alfalfa plants. They have also given silkworms spider genes to see if they would make ready-spun transgenic silk fibres. The resulting silk was a chimera: a bit spider silky, but mostly like silkworm silk. Since then, improvements to the genetic modification process have resulted in much stronger artificial silk that is increasingly like spider silk and less like silkworm silk. But perhaps the most famous and hair-brained attempt at splicing spider genes into another organism was the ambitious yet short-lived attempt at breeding spider-goats. Seriously.

Can you milk silk from a goat?

An initial problem when putting spider genes into *E. coli* was that the silk proteins created were much shorter than natural spider silk proteins. Since *E. coli* have a small genome, the scientists could only insert short fragments of spider genes into them, and the *E. coli* would still only produce fragments of these. One obvious next step was to find a bigger host for the spider gene that could produce longer protein molecules and do it in much larger quantities. This idea piqued the interest of two out-of-the-box thinkers – Professors Randy Lewis and Jeffrey Turner – as well as our good old friend the US military. As we've seen many times before in human history, things tend to get a little crazy when the US military gets involved, and this story is no exception.

Turner had previously researched genetically modified animal mammary gland cells to essentially custom design different types of milk. Years earlier he had left a research career to apply this knowledge commercially and he became the CEO of Canadian biotechnology company Nexia. With backing from the US military, scientists from Nexia collaborated with Lewis – a pioneer in spider silk research – to see if they could put spider genes into mammals. Don't worry, they didn't just jump straight into splicing spider genes into live goats ... yet. They started with cell cultures – tissues grown in a lab from bovine mammary glands and baby hamster kidney cells. Yes, you read that right, baby hamster kidney cells. Lo and behold, as the cells started growing and multiplying in the lab, they started making silk proteins that could be spun into silk fibres.

With a proof-of-concept study showing that genetically modified mammary cells could make silk proteins, the obvious next step was to shoot for the stars and breed their very own line of animals that they could milk for their artificial spider silk. The first step was to trial the method in lab mice, but this was only a step towards a larger goal, and in 1999 the world was introduced to the first pair of genetically engineered spider-goats, named Sugar and Spice. It wasn't long before there was an entire herd of silk-producing pygmy goats running around. The milk these spider-goats produced was just like any other goats' milk – with one important difference – it now had spider silk proteins in it. The goats could be milked just as you would any other goat. Then the silk proteins could be isolated from all the other milky bits before being spun into artificial silk fibres. All was going well until 2009 when Nexia filed for bankruptcy and the dreams of a flourishing spider-goat industry came to an end. With financial backing from the US military, Lewis continued researching spider-goats and their silk for a short time after the closure of Nexia, but has since turned his efforts to perfecting transgenic silkworm silk. In the end Turner and Lewis faced the same problems as other entrepreneurs before them: the yield and quality of spider-goat silk was simply not good enough to make up for the costs and effort involved.

After Nexia closed its doors, Sugar and Spice, the original pair of spider-goats, were sold to the Canada Agriculture and Food Museum and put on public display. While this museum has numerous live animal exhibits, coming face-to-face with genetically engineered spider-goats proved to be confronting for many museum visitors and activist groups. Despite the fact that Sugar and Spice looked and behaved like regular goats, and there were no plans to breed any more spider-goats, the public backlash led to them being removed from display in 2013.

Commercialising transgenic spider silks

Despite huge leaps in synthetic biology, the procedures involved in making artificial silk are complex enough that it still remains logistically and cost-prohibitive. We still haven't quite reached the industrial holy grail of making fast, cheap, super-strong silk fibres. And while we can make some pretty darn impressive bio-inspired materials, there is still something special about natural spider silk that we can't quite match.

We can't replicate the nanoscale processes that happen when spiders intricately combine different proteins from different glands and pull them out through their spinnerets. There is a certain *je ne sais quoi* to the process that that makes a superior material to any artificial process we have ever been able to develop. Attempting to precisely replicate spider silk and all its varied qualities is probably misguided and not all that practical. Rather, the approach of artificial silk producers is now to create 'bio-inspired' fibres, which are based on what we have learnt from studying spider silks, but are unique and have their own special material properties.

There a few companies that have taken these developments in artificial silk beyond the laboratory and into the commercial sector. Spiber (Japan) and AMSilk (Germany) both specialise in using *E. coli* to grow artificial silk, whereas Kraig Biocraft Laboratories (USA) relies on genetically modified silkworms, and Bolt Threads (USA) uses a trademarked strain of yeast. In 2019 Spiber collaborated with clothing company The North Face to create the first 'mass-manufactured' jacket made from artificial spider silk. When I say 'mass-manufactured', that's a bit relative. Initially, there were only 50 jackets made and you had to go into a lottery to win the privilege of spending US$1400 to own one. To be fair, this is quite a lot cheaper than the priceless garments woven directly from spider silk, but for most of us we will be waiting a long time before we can get our hands on a spider-inspired silk jacket. Spiber is still working towards this goal, with the construction of a new mass-production facility in Thailand, so watch this space.

In comparison, Kraig Biocraft Laboratories (also known as Kraig Labs) is targeting a very different market niche. Its flagship product, Dragon Silk™, was produced on contract for the US military to be built into body armour panels, as a more lightweight and flexible alternative to Kevlar. Recently, Kraig Labs has also established its own mass-production facility in Vietnam. While Kraig Labs' approach relies on transgenic silkworms, it stands on the shoulders of spider-goat giants. Randy Lewis, one of the spider-goat creators, also worked to create transgenic silkworms, which led to the founding of Kraig Labs. Lewis continues to research other military applications of spider silk, including a super-strong fibre that could be released into the water to entangle and seize the propellers of enemy warships.

Commercial interest in artificial spider silks has never been greater and this story will continue to be written. So far in this chapter we have only explored spider silk for its super-strong textile and engineering potential. Yet there is another industry that stands to benefit just as much, if not more, from artificial silk. Medical industries are eyeing off spider silk and investing heavily in artificial silk manufacturing, not because of its bulletproof, warship-stopping potential, but for its more delicate and sensitive side.

Spider silk in ancient and modern medicine

Pliny the Elder was a military commander during the early Roman Empire. While he was busy pillaging his way across the globe, he also took time to smell the flowers and was an influential naturalist and philosopher. Thankfully for us he was also a prolific writer. Much of what we know about ancient Rome comes from his extensive descriptions of the ancient world. His greatest and only surviving work, *Naturalis Historia*, is often described as the world's first encyclopedia. In its 37 volumes, Pliny describes everything from art and history to astronomy and zoology. The sections covering medicine collate many of the ancient remedies of the time, which often seem to involve bathing in different varieties of urine, but a handful also involve using spider silk and its apparently miraculous qualities. Most famously, spider silk was used by the ancient Romans as a dressing for wounds and, according to Pliny, spider silk was also an effective remedy for nosebleeds, bruising, fevers, toothache, earache, skull fractures and eye fluxes (whatever the hell they are). Thankfully, urine bathing and many other ancient remedies have fallen out of fashion, but when it comes to spider silk the ancient Romans may have been onto something. Our medical interest in spider silk continues today and in recent years has undergone a resurgence. With advances in bioengineering technologies, spider silk holds genuine promise as a basis for new medical treatments.

Just as spider silk's material properties get engineers chomping at the bit, it has also piqued the interest of medical researchers for its sterility. Think about all the spider webs you have seen in your backyard or in the wild. Have you ever seen a spider web rotting? They definitely start to go brittle and degrade, but people rarely ever see spider webs coated in mould or bacteria. This is incredibly strange. Spider webs are built from nutrient-rich proteins that should make excellent food for decomposing bacteria and fungi. Furthermore, silk persists in places that should be perfect for decomposition – silk lines the burrows of funnel-webs built into moist soil, and webs will stay intact for weeks, littered with dead leaves and insect corpses but the silk seems immune to infection.

This has led people to assume that spider silk must have some sort of anti-bacterial property, but it doesn't seem to be the case. Silkworm silk contains a gum called sericin that kills bacteria, but there is nothing in or on spider silk that actively kills bacteria. Recent evidence suggests that spider silk may actually be 'bacteriostatic', which means that it might not kill bacteria, but can halt bacterial growth. How it does this isn't clear but it appears to have something to do with limiting bacterial access to nitrogen. Nitrogen is an important chemical for bacteria, and while spider silk is rich in nitrogen, something about the way the silk is constructed means that bacteria can't access the nitrogen inside. So, when bacteria come into contact with spider silk, they aren't killed, they just go to sleep, so to speak.

This bacteria-halting property may have been part of the secret behind why the ancient Romans found silk to be an effective wound covering. There are other historical mentions of matted spider silk being used to stop bleeding, including reports of French soldiers in the 1300s carrying thick packets of web into battle. This potential is still being investigated today. Recent studies have trialled dressings made from spider silk on wounds in rats and sheep. In both cases, wounds appeared to heal better when using silk dressings compared to traditional dressings. Other researchers have investigated whether spider silk can be used as sutures; or as us plebs call them, stitches. Using a miniature rope-making machine, scientists wove strands of silk taken directly from golden orb-web spiders into threads up to 360 fibres thick. These threads matched conventional suture thread in thickness, but had increased strength and were able to maintain their strength even after repeated stretching.

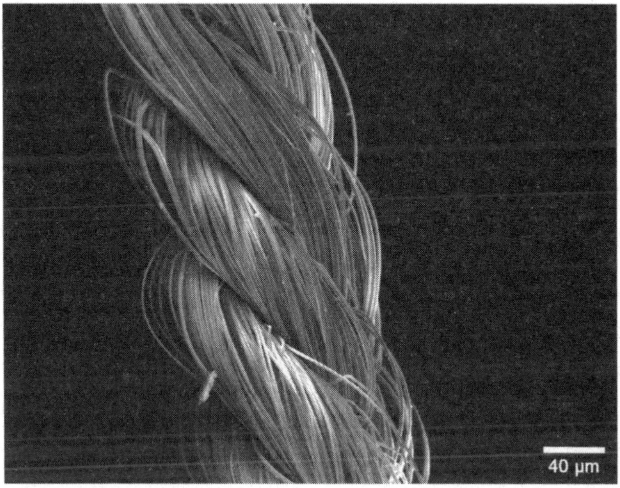

Trichonephila clavipes spider silk woven into experimental suture threads
Hennecke et al. 2013, 'Bundles of spider silk, braided into sutures, resist basic cyclic tests: Potential use for flexor tendon repair'. *PLoS One 8*, 1–10.

If silk's strength and bacteriostatic properties weren't impressive enough, it gets better. For some reason spider silk is incredibly biocompatible; that is, it doesn't provoke much of an immune response when on our skin or implanted inside our bodies. This obviously adds to silk's attractiveness as a material for wound dressings and sutures, and has also sent medical researchers looking for ways it can be used in surgical implants. One promising possibility is using spider silk to help regenerate damaged nerve tissue. Where a serious injury severs nerve connections, people can be at risk of severe disability, including the loss of feeling and function in their limbs. Repairing damaged nerves is incredibly difficult. Scar tissue can stop nerves growing back properly or lead to the formation of neuromas – painful bundles of malformed nerve tissue. To help nerve cells grow in damaged tissue, scientists have been developing physical scaffolding that nerve cells can grow on, and spider silk is proving to be an ideal material for these scaffolds. A team of scientists from Germany and Switzerland took dragline silk from golden orb-web spiders and, in the laboratory, used the silk as a support structure to grow Schwann cells – the cells that line and insulate our nerve cells. When nerve cells regrow, Schwann cells can provide a conduit that guides the growth, leading to healthy, elongated nerve fibres. The scientists then took this principle and tested it in lab rats to see if a scaffolding of spider silk with lab-grown Schwann cells could be used to regrow nerve tissue. This way, rather than nerve cells being left to find their own way through scar tissue, from one side of a wound to the other, the cells had an organic structural mesh to grow along. Silk successfully helped bridge two centimetre–wide wounds in lab rats, which might not sound like much, but think about having a two centimetre hole in your skin and it suddenly seems pretty traumatic. Now imagine something the size of a rat having the same size injury and it's an enormous wound to heal over. After six months the rats treated with spider silk grafts had well-functioning limbs with healthy and well-organised regenerated nerve fibres. Compared to other artificial fibres used to support nerve regeneration, spider silk appears to promote faster and more orderly nerve growth. There were no signs of any foreign-body immune response towards the spider silk and, since the silk is an organic substance, it eventually breaks down into non-toxic components that can be easily removed or ingested by the immune system. This is in stark contrast to other artificial fibres, which have been shown to break down into highly acidic compounds that end up damaging the very tissues they are supposed to be regenerating. Other studies have shown natural spider silk to be a similarly effective scaffold for growing human skin, bladder and cartilage cells.

Despite successful examples of using pure spider silk in medicine, its feasibility as a broadly used treatment is still affected by the same problems faced by the textile industry. Directly harvesting from live spiders is tricky and expensive, not to mention unpredictable because silk quality can vary depending on environmental conditions and the condition of the individual spiders. Not surprisingly, medical researchers are also looking towards artificial silk as a solution to these problems. Artificial spider silk proteins can be spun into silk fibres that could be ideal for making strong and flexible suture threads. They can also be woven into sheets to be used as bandages and wound dressings, or as a substrate for growing cell cultures. And it also turns out that there are a number of different ways that raw silk proteins can be processed that cause them to form different shapes and structures. This is opening up whole new possibilities for how artificial silk can be used in medicine.

With some clever chemistry, spider silk proteins can form porous foams that have been investigated as a three-dimensional medium for growing human stem cells. Using an emulsion process, silk proteins can be coerced into forming spherical capsules. This process is being touted as the future of drug delivery where chemicals can be delivered into your body in microscopic, safely ingestible, time-delay capsules. We can cast silk proteins into thin films to be used as a solid surface coating. This has been trialled in rats as a surface coating for silicone implants. Rats with silk-coated implants suffered fewer postoperative complications such as inflammation and excessive scarring around the implant. Silk protein films are also being tested for their suitability in wound dressing and skin grafting. Finally, as with all good bioengineering stories, 3D printing gets a mention. Scientists are already experimenting with using spider silk hydrogels to 3D-print custom designed scaffolds for cell growth.

It's early days for the spider silk medical industry, but the amount of investment in research is staggering and it's rapidly delivering promising results. While few spider silk medical products are currently available, people are enthusiastically imagining a future where you can 3D-print a spider silk scaffold, to grow a piece of perfectly sized and shaped cartilage, using cultures of your own cartilage cells to rebuild an arthritic limb without the implant being attacked by your body's immune system. Only time will tell whether this future becomes a reality.

Simple and elegant silk solutions

While the French spider silk revolution didn't amount to much, and the much-touted synthetic spider silk industry is still finding its feet, it seems the best uses of spider silk to date have been some of the simplest. Like building an improvised wall out of spider webs to protect yourself from escaped fruit flies, perhaps. Up until World War II spider silk was used to create ultra-fine 'crosshairs' by stretching thin strands across the glass lenses of gun-sights, microscopes and telescopes. There are reports of bundled spider silk being used as fishing lures by Melanesian people of the Soloman Islands, and the

Gulngay people of northern Australia. The people of south-west Malakulu in Vanuatu make their own spider silk textiles, not through weaving but through matting and compressing together many different spider webs into thick sheets; much like wool felting. These can be moulded into undulating shapes, making sculptured pieces of clothing for example. The most well-known of these are ceremonial headdresses worn by high-ranking individuals or during initiation ceremonies.

This chapter, and most of the effort into using spider silk as a material, has been focused on finding very practical uses for it in textiles, medicine and engineering. But silk also has the power to inspire beautiful works of art. Dr Shigeyoshi Osaki is a Japanese scientist who has studied the physics and chemistry behind golden orb-web spiders' super-strong silk, though he is perhaps most famous for another work that took spider silk in a whole new direction. Like many before him, Dr Osaki drew silk from the abdomens of large *Nephila pilipes* spiders and spun it together into thick fibres. Rather than weaving these fibres into a fabric, he twisted bundles of them together to make spider silk violin strings. Professional musicians invited to play using Osaki's spider silk violin strings noted their playability and their unique tone and feel. The tightly packed and super-strong strings are apparently more similar to traditional violin strings made from animal gut than the hard steel and nylon of contemporary strings.

Tomás Saraceno is an Argentine artist who continues to be inspired by spiders and their webs. Among his many spider-inspired works, he created a series of installations that aimed to transform spider webs into musical instruments. In one of these works, *Arachnid Orchestra. Jam Sessions*, live orb-web spiders were placed on display. The vibrations of the spider webs, which are imperceptible to humans, were recorded using a laser Doppler vibrometer, and then modulated and amplified into a pitch that could be heard by humans. Musicians and performers were then invited to perform improvised musical pieces in the space, in response to the sounds being generated by the spider. The result was, essentially, a piece of music composed in real time by both the spider *and* the musicians.

That silk has inspired the creation of music makes complete sense. While we have spent all this time talking about its strength and prey capture qualities, it is equally important to spiders because of its acoustic properties. It transmits sound and signals through vibrations. We have already explored in this book how spiders can use their silk to listen, and in the next chapter we will discover that they can also use it to play songs.

Chapter 7

BIRTH, SEX, THEN DEATH

Toxeus magnus is a small jumping spider that, in some ways, is like many other jumping spiders – it's small and black, it looks a bit like an ant, it's adorable, but overall is pretty inconspicuous. In 2018, a team of scientists from China discovered a surprising detail about *T. magnus* that separates this spider from all other spider species. They lactate. This discovery changed forever the way we see spiders, and maybe even how we see our place in the world.

Toxeus magnus females build a small nest out of silk. They lay their eggs inside and when the spiderlings hatch out, they stay inside the nest and live together with their mother. This on its own is not unusual. While most spiders are solitary creatures, there are many examples of spider mothers that live together with their offspring. Baby *T. magnus* spiders never leave the nest, which means they never go out and catch their own food. This was a bit of a puzzle for the scientists because the spiderlings stayed in the nest for over a month, and continued to grow in size almost until they were fully mature. One potential explanation was that the mother was bringing food back to the nest to share with her spiderlings, but when the team of scientists looked inside the nests to see what was happening, they discovered something very different going on. The spiderlings weren't feeding on prey, they were drinking milk – spider milk. The mother spider secretes a nutritious liquid from a small gland on the underside of her abdomen. Often, she will place little droplets of milk on the walls of the silk nest and the spiderlings will feed from the droplets. Other times, the spiderlings will suckle straight from the mother. As far as we know, *T. magnus* is the world's only known lactating spider.

We think that mammals are pretty great, probably because we are mammals. Lactation is something that characterises mammals and we tend to think it's special. And yet, here we have a spider that lactates to feed her own children in a way that is astoundingly similar to mammals. Just like us, there is a period when the babies will feed solely on milk. Then there is a weaning period when the babies will feed on a mixture of milk and solids. After that, the spiderlings are old enough to go out and catch their own prey. Altogether, the spiderlings stay inside the nest feeding on milk for about 40 days. This is a huge amount of time considering that the spiders mature at around 60 days, and larger animals like mice only breastfeed for about 40 days.

Toxeus magnus are not alone in the list of amazing spider mothers. Many species will devote themselves, and in some cases give their lives, to protecting and providing for their offspring. In this chapter we'll meet a score of amazing mothers, but first I want to give a small mention to spider fathers. Let's face it, they're not great dads. Males don't really hang around for very long after mating. To be fair though, spider dads have a complicated relationship with spider mothers and definitely get the short end of the stick when it comes to the mating game.

Sex, songs and suicide

If you have gone for a wander around the garden and stumbled across an orb-web spider sitting in the middle of its web, chances are that spider is a female. Among orb-web spiders, males are usually a fraction of the size of the females. While females will spend most of their time on the web, males are more active and have to go in search mates. Cast your mind back to the beginning of this book where we envisaged a large spider sitting in the centre of her web listening out for the sound of prey. Now imagine what it must be like for the small male that finds a potential mate sitting in the middle of a giant trap.

If the female spider isn't interested in mating, or mistakes this male for something else, she can quite easily eat him. So how does the male spider make it all the way across the web without becoming dinner? He does what many other male animals do – he plays her a song. When male *Argiope keyserlingi* spiders step onto a female's web, they quickly shake their legs and shudder the web, creating vibrations that let the female know he is there. You might be thinking that this doesn't count as song, because they aren't using vocal cords, like a bird. But as far as scientists are concerned there isn't really a huge distinction between a male bird vocalising to nearby females and a male spider strumming silk. Scientists refer to behaviours like this as 'vibrational songs'. Just as some animals transmit vibrations from vocal cords through the air, spiders transmit vibrations from their legs through silk threads. Spiders don't have anything that we would call an ear, but they often do have special organs on their legs that are sensitive to vibrations. This lets them detect all kinds of signals and cues being transmitted through silk, the ground, the air, and even through water. When a female *A. keyserlingi* hears the song of an approaching male, she can use this information to decide what to do with him. Females are less likely to attack approaching males who shudder webs on approach, showing that these songs play some role in winning the female's favour.

Tropical wandering spiders (genus *Cupiennius*) don't court on the web, they live on broad-leaved bromeliad plants where males will woo females by drumming a tune on the surface of a leaf. Using his pedipalps (a small pair of legs near his head) and his abdomen, a male spider will quickly tap the leaf surface. Just as birds have different phrases and motifs in their song repertoires, these spiders can mix it up with up to 50 different vibrational syllables in their courtship songs.

Male tarantulas are faced with the logistical problem of convincing females to come out of their burrows to mate. While male tarantulas will leave their hiding places to go looking for a mate, females rarely ever leave theirs, and can sometimes spend their entire lives living in the same burrow. Males can detect females using the scent of pheromones on the silk surrounding their burrows. When a male wants to announce his presence, he will vibrate his whole body and tap on the silk and ground around the burrow using his pedipalps and walking legs to send signals that vibrate through the soil. If the female is interested in him, she will start tapping her own song back to him. In some species the female even has a 'bugger off' song she can play, warning males that they should make tracks, or else.

Some male spiders don't just play songs to woo females, they also dance. Colourful peacock spiders are the most famous but they aren't the only ones who use their dance moves to impress females. Wolf spiders (Lycosidae) will drum songs with their pedipalps and also wave their long forelegs, showing off contrasting black and white tufts of hair.

Where song and dance aren't enough to win over a female, males can resort to gift giving. Male nursery web spiders (*Pisaura mirabilis*) will approach a female, holding out a prey item they have recently caught – like a small fly or a cricket – as an offering to her. Sometimes they even gift-wrap the present with a couple of layers of bright white silk. If the female spider accepts, she will take the gift from the male and start eating it while he moves into position and gets down to business. Gifts serve multiple purposes: they can make females more receptive to mating, they keep the female well fed and less likely to cannibalise her mate, and rather conveniently keep her mouthparts busy so that she is focused on eating the gift and not the male. Males don't always keep their side of the bargain and some sneaky males will cheat the system by tricking females with a gift-wrapped piece of bark, or an empty insect exoskeleton.

Being a sneaky male is an effective strategy, evolutionarily speaking of course. In many species, males won't bother with courtship or gifts at all. If you see the web of a golden orb-web spider, take a minute to appreciate the majesty of the female in the centre, then look around the web to see if you can find any tiny males hanging around. The bigger the female, the more males there are likely to be on the web. These little guys are all waiting for their perfect opportunity to sneak in and mate with the giant female – such as when she is distracted by food. I witnessed this recently when I found an enormous golden orb-web spider (*Trichonephila edulis*) one night with a small male sitting just opposite her on the other side of her web.

She was perfectly positioned at about head height so I got my camera out to take a few photos. Luckily, just as I started taking photos, a large moth landed in the web. The female jumped into action. She ran over to the moth and started wrapping it with silk. Right on cue, the male made his move. He ran to the female, climbed right over her back and sneaked in for a quickie while she was distracted by her prey. When he was finished, he casually climbed off her again and went back to loitering around the edge of the web.

Before males can mate, they need to have done a little preparatory work to get their sperm ready for action. We talk about spiders having eight legs, but if we're going to be pedantic they actually have a few more. They have eight walking legs, plus some additional appendages like a very useful pair of tiny legs right underneath their heads called pedipalps. All spiders have pedipalps but they are more pronounced in males and are crucial for mating as the male spiders use them to transfer sperm. First, the male spider will build a little web called a sperm web. He will squirt out a droplet of sperm from his genitals (a small hole under the abdomen) onto the sperm web. Then he reaches down with one of his pedipalps and picks up the droplet of sperm. Picture the pedipalps like a pair of slender arms with boxing gloves on. The boxing glove bit has an inbuilt tubing system that sucks up the sperm droplet. When it's time to mate, the male reaches out to the female's genitals, which are located on the underside of her abdomen. Once the male has found the female's genital opening, he inseminates the female by inserting part of the pedipalp and pumping sperm from the tubing system.

Sometimes he will then remove his pedipalp, but strangely many spiders decide to leave them there. In many species the male will actually detach part of his pedipalp, leaving it inside the female. This is, rather romantically, called a 'genital plug'. Many animals have evolved ways of physically plugging up a female's genitals after mating as a way of preventing other males from mating with her. Some spiders have refined this technique by secreting blobs of hard or waxy substances into the female's genitals, but other spiders stick with the simple method of ripping part of their legs off. These plugs don't stay in the female permanently, but may stay there long enough to ensure that particular male's sperm is used to fertilise her current clutch of eggs.

When a male finds a female and begins mating, the danger still isn't over as he could potentially be cannibalised at any point. So, the mating process often involves a tangle of waving legs and some ritualised behaviours. Tasmanian cave spider (*Hickmania troglodytes*) males have a special crook in their second pair of legs that fits precisely around the female's head. He uses this during mating to keep the female's fangs at bay while he gingerly reaches out with his long pedipalps. Tarantulas mate face-to-face, with the female arching up and the male ducking under to reach in with his pedipalp. This places the male in the unfortunate position of having his head directly underneath the female's poised and ready fangs. Males also have specialised forelegs that clasp onto the female's fangs, keeping them at arm's length. They continue their vibratory courtship repertoire during mating and will tap the female's legs and underside with their legs and pedipalps, presumably as a way to keep the female relaxed.

Once mating is over, there is no guarantee that the male will get away safely. Females can still decide to turn their mate into a tasty meal. It's no surprise then that males don't often hang around for post-coital chitchat and, if they can, will make a run for it as soon as the deed is done. *Philoponella prominens* has made an artform of the dash for the door: males have evolved a special technique to catapult away from females at high speeds the instant that mating is over. While mating, the males will curl their front legs over and hold them between themselves and the female. In this position, hydraulic pressure builds up inside their legs. When the male is finished, he relaxes his leg muscles, and the release of hydraulic pressure flings his legs outwards, pushing against the female. With this split-second manoeuvre, he rockets away from the female with nary a 'thank you ma'am', at speeds of up to 88 centimetres per second.

For some male spiders, sexual cannibalism seems to be the unfortunate outcome of a battle of speed and wits. Other spiders seem to have resigned themselves to their fates and cannibalism has become a formal part of the mating ritual. Male Australian red-backs (*Latrodectus hasselti*) literally somersault into females' mouths when they have finished mating. When red-backs mate, the female usually hangs belly up on her web. The much smaller male sits around her middle, above her genital opening and facing the same direction. To begin the proceedings, he starts tapping around her genitals with his pedipalps. Eventually he inserts a pedipalp, waits a few seconds, and then does something that seems like a pretty stupid idea. The male keeps his pedipalp inserted and flips his whole body over, placing his abdomen right up against the female's mouthparts. There are no prizes for guessing what happens next. She eats him. The head-end of the male can still be busy transferring sperm while the back end is being digested. Sometimes the female is nice enough to wait for the male to finish up with one pedipalp and let him have another go with the second. He will detach, reorient himself, and insert his second pedipalp before doing yet another somersault right into the female's mouth. Even if the female starts eating the male during his first mating attempt, this doesn't deter him from trying again. He will drag his liquefying body around to where he started, do another somersault, and the female can then finish the job while he finishes his.

Dark fishing spiders (*Dolomedes tenebrosus*) win the award for being masters of spider suicide, if that's even worth bragging about. Males only mate once and they are always cannibalised after. Their mating behaviour seems like any other spider's at first. The male signals his intentions to the female with some flashy arm waving. If she is receptive, he will approach, crawl onto her underside and insert his pedipalp. After a few minutes the copulation is over and the male just ... well ... curls up and dies. Literally. He finishes mating, his legs curl up underneath him, and he just goes limp, still dangling from the underside of the female. Somehow, maybe using some Jedi mind trick, he just spontaneously dies. She is then free to grab him and eat him whenever she likes.

If some spiders can evolve crazy evasive manoeuvres to avoid cannibalism, why would other spiders just give themselves up to it? It turns out that there are actually some benefits to being cannibalised. Remember how some spiders will rip their little reproductive legs off and leave them inside the female? Well, these little legs don't grow back so, if we do some simple arithmetic, we'll quickly realise that a male who has done this twice can't ever mate again. If so, then what's the point in living, right? St Andrew's Cross (*Argiope keyserlingi*) males can only use each pedipalp once. After their first mating, the males stand about a 50 per cent chance of being cannibalised, but after their second mating they are cannibalised 100 per cent of the time.

Even if males don't lose their pedipalps when mating, there are still some evolutionary benefits to being cannibalised. Male red-back spiders who volunteer themselves up for cannibalism mate for longer, which gives them more time to transfer sperm into the female. This then leads to a greater likelihood that his sperm will be used to fertilise her eggs. In the spontaneously-dying dark fishing spiders, females who cannibalise males lay almost twice as many eggs that hatch larger offspring compared to females who didn't cannibalise males. So, the male's sacrifice could lead to him fathering more and better-quality offspring.

It might sound weird talking about sperm being *used* to fertilize eggs, as if the female has some choice over what sperm fertilizes what eggs. Actually, she does. A female spider has special organs where she can store sperm from multiple matings before they are used to fertilise her eggs. This gives her a certain amount of control over which sperm she uses. How she does this is can be pretty complex, and for many species we don't know precisely how they do it. But whether its conscious, or instinctive, or some kind of spider Kegel exercise, we know a female can exert *some* level of control over the paternity of her eggs. For spiders who lay clutches of several hundred eggs, it's unlikely that all of those eggs will have been fertilised by just one male. Anything that a male can do to increase the chances that his sperm is used, like plugging up the female with discarded appendages, or mating for longer so that there is more of his sperm inside her, will hopefully increase the number of offspring he passes on to the next generation.

With all those crazy rituals over and done with, the male has hopefully fertilised a few eggs and that's his job done, whether he has survived to tell the tale or not. From here on in, it's all down to the mother to give those offspring the best chance of survival.

Raising good spiderlings

Spiders famously lay egg clutches with many hundreds of spiderlings. Large golden orb-web spiders can lay clutches of over 3000 spiderlings. And they can do this multiple times in their lifespan. Spiderlings are such small, vulnerable creatures that laying as many eggs as possible is a good strategy for ensuring at least some of them survive. And while these spiderlings are developing inside their eggs they are treated with the utmost care. Spider mothers weave thick silk egg cases that caress and protect their eggs. Usually this involves the mother building a silk disc onto which the eggs are laid and then covered with another layer of dense silk, creating a protective pillow. Silk egg sacs protect the delicate eggs inside from the elements and from physical disturbance, as well as from parasites and predators. In some species, the egg cases look like hard-shelled balls, while in others they can be a fluffy cotton-like ellipse. Some species adhere their egg sacs flat against a surface while others are camouflaged in the environment. Orb-web spiders from Costa Rica have been observed incorporating dead leaves into their egg cases. The mothers pull a dead leaf up from the forest floor using lines of silk, then lay their eggs into a silk mat placed on the surface of the leaf, next wrapping lines of silk around the leaf, causing it to roll around the silk egg case. Once the curled leaf is sealed tight it is lowered back to the forest floor and cut free. Tasmanian cave spiders construct elaborate multi-layered egg cases that hang on a stalk attached to the cave roof. The outer shell is a dense layer of bright white silk that is lined with an inner layer of sparse insulating silk which protects the eggs that are held suspended inside a large globule of fluid.

Mother spiders will do whatever it takes to protect their egg cases from predators and parasites. Tarantula mothers keep their egg sacs safe inside their burrows. Some huntsman spider mothers carry their disc-shaped egg sac around with them under their bellies. Wolf spider mothers attach their egg sac to their spinnerets and carry it around behind them, whereas nursery web spiders carry their eggs in front by holding the egg sac in their jaws. Orb-web spider mothers will place their egg cases directly on their prey capture web, where they can stay nearby and fend off any attackers. Some tarantulas will weave barbed hairs from their own skin into the surface of their egg cases as a defence against ants and parasites.

Cupiennius wandering spiders take refuge inside bromeliad plants where deep pools of water well at the base of the leaves. When a female lays eggs, she builds a thick membrane of silk across the reservoir and hides behind it, holding her egg sac underneath her. If she is disturbed, she will dive into the reservoir of water and take her eggs with her. She can stay submerged for over an hour without any damage to the egg sac or spiderlings.

Spiders have been defending their eggs for millennia. In 2021 scientists discovered fossil spiders from Myanmar preserved in amber that appear to be a snapshot in time of a mother spider guarding an egg sac full of spiderlings. And although we discounted spider fathers as active participants a while ago, there are species like *Manogea porracea* that like to remind us it's not always that simple. This species from Brazil has a very contemporary way of life. The male will build his own small web directly above the female web, and the eggs are laid on strands of silk that stretch between the two webs. It's almost like the parents have separated and dad has moved out into his own apartment next door so that they can share custody of the kids. Both spider parents aggressively defend the eggs from predators, but by the end of the reproductive season the dads are more likely to be left looking after the kids. In field studies of this spider the males tended to have longer lifespans than females. So, it was more common for males to be left looking after the eggs once the female was no longer there. Even without the mother present, the male will still stay around repairing the web, defending the eggs, and even cleaning the eggs after rain.

But parental care goes much further than just defending eggs and spiderlings. Again, this is where spider mothers go above and beyond to give their babies the best chance of survival. When it is time for the spiderlings to emerge, either the mother spider will chew a small hole in the egg sac or the spiderlings will chew their own way out. Some spiderlings will live together with their mother for a short time. Among spiders that live in burrows or crevices, like tarantulas and huntsman spiders, the spiderlings will stay inside the refuge for weeks, huddled around their protective mother. The spiderlings don't need to leave to hunt; their mother will share food by regurgitating some of the prey she has caught, like a mother bird returning to her nest with food for her chicks.

Wolf spider mothers famously carry their clutch of spiderlings around with them on their backs. As the spiderlings emerge from their egg sac, the mother sits motionless while her spiderlings climb onto her abdomen and form a dense crawling layer of baby spiders. With her brood accounted for, the mother wolf spider can move around, seek shelter and avoid predators. If the spiderlings get knocked off, they can crawl back onto their mother by climbing up her legs or up the dragline silk she leaves behind her. When the mother spider finds some water, her spiderlings will climb down her legs and have a drink before returning to the safety of her abdomen.

Black lace-weaver (*Amaurobius ferox*) mothers and their offspring live together for a short time in a silk-lined retreat hidden under rocks or in crevices. When her eggs hatch, the babies cluster underneath her. About a day after hatching, the mother will lay another clutch of eggs. These eggs won't ever hatch into offspring. This is a special batch of eggs specifically laid as food for her babies. The mother and spiderlings continue living together for a few days more until the mother makes another, more grand, offering of food to her babies. She offers up herself. When the spiderlings are about a week old, the mother becomes more active and starts pulling at the silk lining of their home. She starts tapping her children with her legs and drumming on the ground with her pedipalps. This seems to be some sort of elaborate ritual leading up to her death. Her spiderlings eventually become more active, moving around their home and finally swarming onto their mother in a sudden wave. The mother goes still and the spiderlings feed. She is completely consumed within a few hours. Spiderlings who eat their mothers more than double their weight, and get a developmental head start leading to better survival chances later on.

Black lace-weavers are just one of the many spider mothers who give themselves up as a meal to their offspring. There can even be an element of programmed suicide that makes the mother's death inevitable. The desert spider (*Stegodyphus lineatus*) begins to digest her own insides in preparation for her spiderlings' maternal feast. Once she has laid eggs, the mother no longer feeds herself and her stomach tissues begin to dissolve. After a few days, the rest of her abdominal organs break down and her insides turn to liquid. By the time the spiderlings are about nine days old they have fully developed mouthparts and are ready to start nibbling away at their conveniently liquefied mother.

In most cases spiderlings will depart from their brood and away from their mother not long after they are born. As they grow, they need more and more food, and their own siblings start to look like rather appetising meals. Once they leave the brood, they live a solitary life fending for themselves. But, as these spider stories prove again and again, there are always exceptions to the rules, and there are a small number of spider species that manage to withhold their cannibalistic tendencies well enough to live together in family groups.

Family life

In the tropical rainforests of South and Central America there are spider webs that stretch across trees, smothering the understory in a shimmering white blanket of silk. These are the colony webs of a spider that has evolved to live in large social groups. *Anelosimus eximius* spiders can live in groups of tens of thousands, where they cooperate to build their enormous webs. Living inside these webs are spiders of all ages and sexes, with multiple females laying and rearing egg clutches. In this colony, the spiders cooperate to hunt prey; taking down much larger insects than any individual could manage alone. When an unfortunate insect lands on the colony web, its vibrations will alert any nearby spiders. The spiders attack in a pack, biting the prey and holding it down. They even coordinate their approach towards the prey so that the vibrations caused by their footsteps don't drown out the vibrations of the prey and the group can all successfully locate the prey in the web. Together they envenomate the prey and tie it down with sticky silk. When the prey is immobilised, the spiders cut it free from its bonds and carry it away, where it will be fed on by many.

Unlike *Anelosimus* colonies, which have many females laying and rearing clutches in a single colony, many group-living spiders live in a single matriarchal family group. *Australomisidia* crab spiders live together in nests built from gum tree leaves woven together with silk. These nests begin as many other spider nests do: from a sole female who has left her maternal nest and begins her own. She weaves together a small clump of leaves and lays a clutch of eggs inside. The nest doesn't stay small though. As the spiderlings hatch out and grow, they all contribute to expanding the nest. They add more leaves and more silk, creating a complex labyrinth. Eventually the matriarch will offer herself up as food for her growing spiderlings, and the children continue living in the nest without their mother until they are mature.

Social huntsman spiders don't build a colony web. They live together in small family groups hidden behind tree bark or under rocks. These are matriarchal groups with one mother and her offspring from several clutches that overlap in ages. Her children will hang around the family home until they get too big and need to move out. At dusk the mother spider and some of the older children will leave the retreat to hunt before returning at dawn. The mother huntsman will share some of her food with the smaller spiderlings back at home by regurgitating predigested prey. She doesn't feed her older children though; they are big enough to hunt and feed themselves.

Stegodyphus spiders incorporate leaves, twigs and other debris into their colony webs, which are stretched between two branches. The result is a dense mass of plant matter suspended in midair that can contain hundreds of spiders. Mothers will lay multiple egg sacs, so families contain juveniles of varying ages. In the species *Stegodyphus dumicola* the father stays with the colony to guard the female from any other intruding males, and also shares the responsibility of guarding the egg sac.

Biologists like spending inordinate amounts of time classifying different types of sociality, depending on how different animal colonies are structured. They have come up with all manner of fancy category names like parasocial, subsocial, or eusocial, to rank different levels of sociality. Perhaps it's because we're social animals that we are fascinated by other animal groups and how they work, and find it necessary to spend hours splitting hairs and trying to rank other animals on some conceptual sociality ladder. Humans are far from being at the top of this social ladder. The winners of that competition are social insects like ants, bees and termites; they form entire colonies of cloned daughters who work to support a single queen who is responsible for all the reproduction. Where individuals exist purely to raise and protect another individual's offspring is where we perceive the pinnacle of sociality to lie. Spiders don't quite reach these lofty heights of sociality. Often social groups consist of mothers caring for daughters, and the main benefits for spiders seem to be shared foraging and defence. The closest spiders seem to come to sharing responsibility for brood care are in the enormous colonies of *Anelosimus eximius*. These groups are mostly female, with large numbers of mature females present. Yet there are relatively few clutches of eggs for the number of females present, which suggests that only a few of them are reproducing, and the rest are there to help feed and defend the colony of other spiders' offspring.

Death

For most spiders, death comes through the agency of one of the many other predators out there in the wild. In temperate places where winter is cold, many spider life cycles begin and end with the passing of the seasons. Eggs hatch as the weather warms in spring. The spiders mature and grow over summer just in time to mate and lay eggs in late autumn. The adult spiders die, the eggs overwinter and the cycle continues. In tropical areas, this life cycle isn't as seasonally restrictive, and there are many species that can overwinter and go on to live longer lives. For some this might be a few years; for others its much longer. Tasmanian cave spiders seem to live their lives in slow motion compared to other species. Their mating behaviour can last for many hours, compared to the few minutes or seconds seen in other spiders. Their eggs stay inside the egg sac for up to ten months, and an individual spider can live for around 20 years. But the longest lived of all spiders are the tarantulas. When tarantula spiderlings leave their mother's burrow, they search for a place to build their own. For many, this will be the only burrow they ever live in. Year after year the same tarantula can be found living inside the same burrow. Many species are estimated to live for over 30 years, and the oldest tarantula on record reached 43 years. Her name was '#16'.

Barbara York Main was one of Australia's most prominent arachnologists. From an early age she was fascinated by the spiders she found around her family home, and would later forge a career as a research scientist and natural history writer. In 1974 Barbara began a research project on the *Gaius villosus* trapdoor spiders of North Bungulla Reserve in Western Australia, a tiny remnant of bush precariously surrounded by cleared wheat fields. She meticulously searched a patch of bush to find every trapdoor spider burrow possible. Each one was sequentially numbered. When Barbara found her sixteenth trapdoor burrow, she did the same as for every other one: took a small metal peg with a numbered disc on it and hammered it into the ground behind the burrow entrance. The spider in this particular burrow was estimated to be about a year old, based on the width of the burrow. Some of the other burrows she found were much larger and probably housed spiders around 20 years old.

Barbara returned to the site twice a year, keeping track of the different burrows and their spiders. Year after year she would find newly built burrows and add more numbered pegs to her surveys. As the spiders grew, their burrows got wider and wider. She watched as juvenile spiders matured into adults. When adults passed away, their burrows fell into disrepair. Each of the spiders found during the original survey in 1974 eventually died, but #16 lived on. Originally, the study was planned to go on for 20 years, but spiders like #16 proved even that ambitious target wouldn't be enough to fully understand how long trapdoor spiders can live. Barbara's regular visits to #16 and the other spiders would continue on for 40 years, until Barbara, by then in her eighties, wasn't able to do field work anymore. The torch was passed on to Leanda Mason, a student of Barbara's, who continued the surveys of North Bungulla Reserve, always making sure to check in on #16. That was until, in 2016, Leanda arrived at #16's burrow to find it looking dishevelled, and with a suspicious looking hole in the doorway. Leanda opened the burrow but #16 was nowhere to be found. The hole in her burrow lid was the telltale sign of a parasitic wasp that had gnawed its way inside. The news was kept quiet for two years until the two researchers formally announced that #16 had died. Headlines around the world mourned the death of #16, the world's oldest spider, at the age of 43. In 2019, three years after she farewelled her long-time friend #16, Barbara York Main passed away, at the age of 90.

Chapter 8

MASTERS OF THE EARTH

Spiders can swim, *and* fly. No really, I'm being serious. There are scuba diving spiders and mountaineering spiders and aeronautical spiders. It might sound like I'm joking, but spiders push the limits of what life is capable of. In this chapter and the next, we'll explore the extreme lives of spiders, the unexpected places we find them, and the feats of exploration they have achieved. One could even describe the incredible feats of spiders as miraculous; after all, they *can* walk on water.

If you were to shrink down to the size of a spider, you'd discover that water is not the free-flowing liquid you know it as. On the micro scale at which spiders live, water is thick and viscous. If you were to press your hand into a droplet of water, your hand would not sink immediately as it does now. The surface of the water would resist and wobble like jelly. Be careful not to push too hard. If you pierce the surface of the water your hand will get stuck and it could take all your strength to pull yourself free from the thick gooey mass.

Fishing spiders (genus *Dolomedes*), also called water spiders or raft spiders, use the viscous properties of water to walk on its surface. Their legs and bodies are covered in a dense coating of water-repellent hairs. These in combination with the spiders' small size and the surface tension of water molecules mean that fishing spiders can walk right across the water's surface. And not just walk either – fishing spiders can run on water, row across it, and even leap up in the air. Where each leg touches down, the water's surface bends, forming small dimples, like human feet on a trampoline.

During the day, fishing spiders seek shelter along the banks of freshwater streams and lakes, but at night they wander out onto the water's surface to hunt. Just to be safe, they tether themselves to the shore with a length of dragline silk to stop themselves floating downstream. Under darkness, fishing spiders cannot use their eyes to detect prey. Instead, they have to listen. In the same way that other spiders will listen to vibrations in their web, fishing spiders listen to vibrations and ripples in the water. Here they can catch aquatic insect larvae, or perhaps unlucky insects that have fallen into the water and can't get out. If they are lucky, they might even come across a small fish or a tadpole.

If a small animal, say a mosquito larva, splashes against the water's surface, those vibrations will ripple outwards. A spider standing patiently on top of the water can detect those ripples, identify what direction they are coming from, and leap forwards to catch it. The similarities between a spider listening to its web, and a spider listening to the water, do not end with their hunting behaviour. Fishing spiders can use water to communicate by making their own ripples on the surface. In the same way that other male spiders sing to females by vibrating their webs, fishing spiders will bounce up and down on the water's surface, sending out ripples to let females know they are nearby. Female fishing spiders release pheromones that float above the water's surface. When a male spider detects these pheromones or the dragline silk of a female, he will follow the scent by rowing across the water's surface or by pulling himself along the dragline. Once close enough, he will begin his vibratory courtship using downward leg thrusts in combination with some elaborate leg waving. The female can then respond with her own song by drumming the water's surface with her pedipalps.

Many different species of spiders walk on water and use it for hunting and locomotion. Some are opportunistic water dwellers that can survive on water if needed, but are primarily landlubbers. Other species, like those in the genus *Dolomedes*, are expert water hunters with special adaptations that allow for mastery of the water's surface. One species takes aquatic living to a whole new level. It's called the diving bell spider, *Argyroneta aquatica*, and it is the only spider that spends its entire life in the water. And it manages to do this, essentially, by taking a miniature scuba tank with it everywhere it goes. Yes, seriously.

Spiders can swim

The diving bell spider's abdomen is covered with a dense layer of fine hairs. When the spider is submerged, these hairs trap air and hold it against the body, forming a thin bubble around the spider's abdomen, which just so happens to be where the spider's lungs are situated. Spiders do not breathe through their mouths like you and me. Instead, they have a small hole on the underside of their abdomens. This hole is the opening to their book lungs – breathing organs shaped like the folded pages of a book. By trapping air around their abdomens, diving bell spiders can take a ready supply of oxygen with them wherever they go. And when the oxygen levels in that bubble start to drop, they can simply swim back to the surface and replenish it with fresh air.

Now if you think miniature spider scuba tanks are incredible, it gets even better. Would you believe me if I told you that diving bell spiders live most of their lives inside tiny submarines made from silk? The name 'diving bell spider' is inspired by the incredible homes that they make for themselves. A diving bell spider will search for an ideal spot, just below the water's surface, where there are submerged twigs or blades of grass. Between these bits of plant, the spider will start to build a dome-shaped web. It will then swim to the surface, fill its little spider scuba tank with air, bring it back and deposit a tiny bubble inside the dome. After a few trips, the bubble in the dome starts to grow bigger and bigger until it is large enough for the spider to crawl inside. And so, just like early explorers descending into the ocean in dome-shaped diving bells, these spiders survive underwater inside a bubble of air trapped beneath their webs.

The spiders continue to make quick visits to and from the water's surface to replenish the air inside and can survive in their bubble indefinitely. This silken diving bell then becomes the spider's home. They set traps for prey by attaching threads of silk from the bell to nearby vegetation. Like a fisherman lazily watching his lines for telltale shudders, the spider can sit patiently inside the bell, listening for the vibrations of ensnared prey. If successful, they will venture out into the water, kill their prey, and bring it back to the bell to finish the meal. A male spider will search for a female inside her home, courting and mating with her inside the bell. Females lay eggs inside the bell, giving their babies a ready supply of oxygen and a safe place to eventually emerge.

Living underwater even helps diving spiders survive frozen winters. In the harsh winters of northern Eurasia, rivers and lakes freeze over. In the cold water underneath the ice, aquatic spiders stay dormant inside their bells. It seems counter-intuitive but underneath the ice is warmer than above. Air temperatures on the surface can drop *below* freezing, but the flowing water under the ice is always *above* freezing. It has to be, otherwise it would be ice, not water.

While many spiders can dive underwater, the diving bell spider has been seen as a bit of an anomaly in the spider world because of its ability to stay underwater for long periods. However, as more spider species are discovered there's a possibility we could find more aquatic spiders right under our noses. In 2016 in remote northern Australia, near the town of Maningrida, scientists discovered a new species of tarantula. Like other tarantulas, it's long lived and spends its life inside soil burrows, rarely coming outside. This would seem completely normal but for the fact that these spiders live on a massive floodplain. Tropical northern Australia is monsoonal, and during the wet season these spider burrows can be submerged for months at a time. When submerged, the tarantula's coating of thick hairs traps a layer of air that covers the spider's entire body. This allows the spider to breathe underwater and presumably is how it lives fully submerged for several months of the year. The spider has been dubbed the Maningrida diving tarantula but it is such a new discovery that it doesn't even have a scientific name yet. There is so little known about it that we will have to wait and find out how this spider makes a life for itself underwater.

Despite their aquatic proficiency, spiders could hardly be called seafaring. Fishing spiders and diving spiders only live in fresh water, and the closest thing we have to marine spiders are a few species that live in the intertidal zone. They take refuge in air bubbles that form under crevices on rocky shores or in hollows at the base of large kelp plants. Here they can stay submerged for weeks at a time. As far as we know, there are no spiders that have ventured any further into the depths of the ocean. Having said that, spiders are famously adept at crossing oceans. Like some sort of eight-legged version of a character from a Jules Verne novel, spiders can explore watery depths in silken diving bells *and* take to the skies on silken balloons.

Spiders can fly

In the 1920s scientists at the US Department of Agriculture decided to undertake a study of something called aeroplankton. You may be more familiar with sea plankton, the almost invisible microscopic invertebrates and assorted animal larvae floating in the ocean. Another mysterious ecosystem of microscopic animals floats above our heads. Back in the 1920s, as the government organisation responsible for understanding things like crop pest migration and airborne disease vectors, the US Department of Agriculture looked to the skies and wondered what small critters might be floating up there. As the first-ever study of aeroplankton, there was no guidebook as to how they should sample insects in the air. The task fell to the assistant entomologist P.A. Glick, who was given the challenge of inventing a new 'method of observing or collecting them while in flight in some form of aircraft'.

I imagine that hidden away in a US government building somewhere are Glick's ageing notebooks, filled with sketches of marvellous insect-collecting flying contraptions; perhaps a giant butterfly net turned into a high-speed parachute, or special harnesses that tether screaming entomologists to the underside of planes, or maybe a hot air balloon covered in double-sided sticky tape. Without these imagined notebooks to tell us otherwise, we must assume that Glick was a much more practical fellow than myself and invented an ingenious collection box that could be strapped to the wings of a plane. The earliest models of these collection boxes were built from wood and aluminium and contained several 'cartridges' – square wire-mesh flyscreens covered in a sticky solution of castor oil and resin. A wire cable ran from each of these cartridges into the cockpit of the plane. Pulling the cable would slide the cartridge out from the box, exposing it to the air and intercepting insects that got stuck to the glue-covered mesh.

The next step was to recruit hot-shot military-trained pilots willing to strap this contraption to the wings of their planes. Remember, this was all taking place in the 1920s. The Wright brothers had only made their history-defining 12-second flight in 1903. The 1920s was the era of beautiful wooden propeller biplanes and open-air cockpits.* Any pilot who undertook this insect collecting mission would have been an aviation pioneer putting their life on the line for the sake of understanding aeroplanktonic biodiversity. The missions were a success: the cartridges worked and, very importantly, nobody died in the

* For the plane-spotters reading this, they used a Curtiss JN6H biplane, a Travel Air 4000 biplane, four de Havilland H1 biplanes and a Stinson-Detroiter SM1 monoplane.

process. Glick proudly boasted in his report that only 'one major accident occurred, when a forced landing resulted in the destruction of the craft and injury to both the pilot (McGinley) and the writer. Such mishaps must be expected in a more or less hazardous undertaking.' I break out in a cold sweat just thinking about the health and safety paperwork were this research to be conducted in the same way today.

The efforts of these aviators were rewarded, with over 30 000 specimens collected and identified. As would be expected, the haul included a lot of flying insects like flies, bees and beetles. But there was one animal that was found conspicuously higher than any other. At around four and a half kilometres above the Earth's surface, higher than any other flying insect was collected, they found a spider. Glick writes in his final report that during the flights he took part in, he witnessed strands of silk floating through the skies. Often, after the planes had landed, the landing struts would be covered in a thick layer of silk threads.

This single spider found floating through the sky was by no means an anomaly. Spiders are masterful flyers. Anyone who has read E.B. White's classic children's book *Charlotte's Web* will remember the final act when (spoiler alert) a warm draft of air passes through the barn taking all but three of Charlotte's 514 spider babies with it. Under the doting gaze of Wilbur the pig, the spiderlings release long threads of silk, say their goodbyes, and balloon away to live their own lives. The sad truth glossed over in this children's book is that the spiderlings were probably dispersing en masse to avoid the inevitable temptation to eat one another.

'Ballooning' is commonly seen in juvenile spiders and it allows them to quickly move long distances and land somewhere new, away from competition with their siblings and parents. To balloon, a small spider will crawl to a high point, like the tip of a branch, and adopt a 'tip-toe' posture, which is perhaps the best description despite the spider's lack of toes. It stands tall by stretching its legs and lifting its abdomen in the air. It will then release threads of silk – sometimes a single strand, sometimes a fan of multiple strands – that hover in the air like a kite. The general idea is that the slightest gust of wind should be enough to catch the spider's silken kite, lift the spider off the ground and send it floating away to somewhere new.* This may conjure an image of a spider briefly floating through the air and falling not far from where it started but don't be fooled, spiders are capable of epic aerial voyages. As you already know, spiders are capable of reaching several kilometres in altitude. They can also soar for kilometres across oceans. Sailors have witnessed spiders and their webs appearing on their boats when they are nowhere near land. Perhaps the greatest spider journey ever documented is an account of a spider that had flown around 3200 kilometres only to land on the inhospitable shores of Antarctica. This happened in the mid-1900s when scientists caught a mature male spider in an insect net while on an expedition in McMurdo Sound. No spiders are known to survive in Antarctica, and the nearest hospitable areas are subantarctic islands south of New Zealand, so they concluded that the spider must have arrived there by air. Because of their ballooning superpowers, spiders are unexpectedly some of the best dispersers of all terrestrial animals, and are often the first colonisers of new environments.

* Hold that thought for a minute. Things aren't always as they seem and ballooning is about to get weird.

In 1883, the Earth changed, quite literally, when the Indonesian island once known as Krakatau (or Krakatoa) exploded in a volcanic eruption so colossal it has been described as the loudest sound in recorded history. The ensuing tsunami may have even affected sea levels on South American coastlines, on the other side of the planet. The eruption obliterated most of the island, leaving only its southern tip, called Rakata. Any living creatures on the island were destroyed with it, and anything remaining on Rakata was quickly incinerated by fast-moving pyroclastic flows. The entire island was effectively wiped clean and resurfaced with molten lava. Six weeks after the eruption, Dutch scientist Roger Verbeek travelled to Rakata only to find that the ground was still too hot to walk on. It wasn't until the following year that Belgian scientist Edmund Cotteau set foot on the island to search for any signs of life. Cotteau was part of a team that set sail from Batavia (now called Jakarta). Along the way, the team bore witness to the aftermath of the great eruption. They spotted a large paddle-steamer sitting in the middle of the jungle, having been picked up and put there by the tsunami. Tracts of forest washed over by the tsunami were now fields of skeletal tree trunks. What were once bustling villages were flattened plains littered with giant coral-covered rocks hurled ashore.

As they approached Rakata, what remained of Krakatau, it seemed to still surge with volcanic energy. Stones fell incessantly down the jagged, rocky slopes, stirring up ash and shrouding the island in dust. Eventually, the team found a small beach on the western side of Rakata where it was safe to approach. Cotteau searched the desolate plains of solidified ash and pumice. There were no signs of life anywhere except for one small exception: a tiny spider. Soon there would be more signs of life as plants and other animals found their way ashore and began to colonise the new island. But, as far as we know, the first coloniser of Krakatau was a spider that we can only assume ballooned there from a nearby island.

If you are finding it hard to imagine that a gentle breeze could lift a spider off the ground and send it soaring away on an ocean-crossing adventure, then you're not alone. The more that scientists studied ballooning, the more they found that things didn't seem to add up. People have consistently reported seeing entire groups of spiders being lifted into the air on days when there don't seem to be any prevailing winds at all, not even a gentle breeze. As a general rule, ballooning tends to occur when wind speeds are low (less than three metres per second), which seems counterintuitive. And it's not only baby spiders that balloon, either. South African *Stegodyphus* spiders have been seen ballooning as full-grown adults when they weigh over a whopping 100 milligrams. That doesn't sound all that impressive when you consider that 100 milligrams is about a tenth of the weight of a paperclip, but for spiders, that's a chunky little beast.

Charles Darwin made his own observations of ballooning spiders. In 1832, during his famous voyage onboard the HMS *Beagle*, he noticed one morning that all the ship's ropes were covered in fine gossamer threads. The ship was 60 nautical miles off the coast of Argentina at the time so it was pretty unlikely that spiders had walked on board to infest the ship. Darwin soon found the likely culprits when he spotted large numbers of tiny red spiders crawling along the ship's railing. As would be expected of one of the greatest naturalists to have ever lived, Darwin watched the spiders very closely. Now and again, he'd observe the same ballooning behaviour described earlier, where each little red spider would lift its abdomen, release a short length of thread from its spinnerets and sail off into the air. Darwin took meticulous notes, giving the times and dates of his observations, the local weather conditions, and step-by-step descriptions of the spiders' behaviour. One thing that he noted would, centuries later, provide a clue as to how spiders actually fly. Darwin wrote that when the spiders took to the air they would 'sail away in a lateral course, but with a rapidity that was quite unaccountable'. If we imagine that ballooning spiders are being lifted into the air by gentle gusts of wind, we might imagine them lifting slowly upwards like miniature kites. But according to Darwin, they didn't do that at all. The spiders went sideways, and they did it fast.

Perhaps the most obvious hint that something strange is happening, is that the silk spiders use to balloon behaves quite strangely. When spiders are getting ready to balloon they can release several silk strands that seem to fan out as they lift up into the air. As you know, silk is sticky. A cluster of sticky silk threads would be expected to tangle together in a great big sticky mess, not fan out. This is easily explained though if we switch for a moment from thinking about ballooning to thinking about balloons. You have probably done the at-home science experiment of getting an inflated balloon and rubbing it against your head to make your hair stand on end. The friction caused by rubbing the balloon against your hair generates a static charge that causes your hair to fan out and stick to the balloon like a magnet. As each strand of your hair now carries the same static charge (i.e. a positive charge), they repel each other like two positive magnet ends and your statically charged hair fans out, while simultaneously being attracted to the balloon, which has a negative charge. This same phenomenon happens with spider silk. When silk is released from a spider's spinnerets, each strand is electrostatically charged and so repels the other strands around it.

The idea that static electricity could aid in spider flight was put forward as early as 1826 by naturalist John Murray, who had a knack for lyrical and fantastical descriptions of natural phenomena. Murray concluded that silk threads did not tangle together 'from their being imbued with similar electricity'. He even claimed to have seen the electric fields surrounding spider silk when, at just the right angle, he 'observed an extraordinary atmosphere or aura round the thread' that he was certain was electric. Murray experimented with spider silk by carefully holding different objects next to ballooning spiders and observing whether the silk was attracted to or repelled by it. From this, he correctly deduced that silk held a negative charge. Understanding that the air in the immediate atmosphere was positively charged, he suggested that the spiders' negatively charged silk would be drawn upwards towards it.

Since then, attempts to understand spider ballooning have mostly approached it as a fluid dynamics question. Scientists have considered possibilities like whether parcels of rising hot air could lift spiders off the ground, or how air turbulence might create eddies and vortices that lift spiders into the air. In the end, they were often left with more questions than answers, like why spiders don't balloon in high winds, or how they reach such incredible speeds at take-off. It wasn't until almost 200 years after Murray's observations that the importance of static electricity was confirmed and our understanding of ballooning took off, so to speak. In 2018, Dr Erica Morley and Professor Daniel Robert at the University of Bristol placed spiders into enclosures so that there were no prevailing wind currents. They then generated an electric field using charged metal plates on the floor of the enclosure. As soon as the electric field was switched on, the spiders adopted the stereotypical 'tip-toe' posture that preceded ballooning, suggesting that the spiders could detect the presence of the electric field. The spiders then started releasing silk and, inside a completely windless chamber, lifted off the ground. Negatively charged spider silk was repelled by the similarly charged metal plates on the floor and drawn towards the positively charged particles in the air. Morley and Robert even found that they could control the spiders' ascent or descent by simply switching the electric field on and off. Spiders don't balloon: they launch on electromagnetic silk rockets.

You may be wondering where spiders in the wild are finding electromagnetic fields to launch from. The answer is, well, everywhere. Particles in the air have an electric charge and pointy objects, like the tips of leaves and branches, tend to affect the electric field charging the air particles around them. This explains why lightning arcs to things like trees not necessarily because they are the highest objects, but because of the strong electric fields around the tips of their branches. So, when a spider walks to the tip of a branch it can detect the strong electric field in the air, and use it to launch itself skywards. Or, as John Murray emphatically put it, the spiders 'mount on high their electric chariot'.* Under certain weather

* The rest of this quote is even more fun as Murray was proposing that spiders took to the skies to wage aerial warfare with insects. He said that spiders mounted their electric chariots 'to thin the ranks of those clouds of the genus *Staphylinus* which if suffered to remain undisturbed, might, for ought we know, increase even to the destruction of us and ours'. Oh, how I wish I was a nineteenth century naturalist.

conditions, the electric fields around projecting objects are stronger, which explains why groups of spiders tend to balloon all at the same time when conditions are optimised for take-off. This finally explains how spiders can launch at such incredible speeds, as the attractive or repellent forces that electric fields generate can be very powerful. These same principles are what propel maglev bullet trains, the fastest trains on Earth: electromagnetic forces lift trains off the ground and push them forward.

The discoveries of Morley and Robert were inspired by an ingenious bit of research from Professor Peter Gorham from the University of Hawaii who, in 2013, realised that he could use mathematical models to test Charles Darwin's observations of ballooning spiders on the HMS *Beagle*. Gorham took information from Darwin's meticulous notes, including the weather conditions and the time, date and location of the observations. He also gathered information about the HMS *Beagle*, like its size, shape, the fact that it had a lightning rod attached to the mainsail, and that its wooden hull would have absorbed salt water from the ocean. Using this information, he built a computer simulation of the electrostatic potential of the ship on the day of Darwin's observations. He found that there would have been strong electric fields around the ship's railings that would have repelled a similarly charged object horizontally away from the ship. This is exactly what Darwin observed in 1832: small spiders on a calm, clear day rapidly launching sideways from the ship's railings.

All this is not to say that wind doesn't play any role in spider ballooning. We don't know much about what happens once a spider is in the air, whether they have any control over their flight or if they are simply at the mercy of the wind and weather conditions. As far as take-off is concerned though, the term spider 'ballooning' might be an understatement as we now know it's way cooler than lighter-than-air travel. Spiders don't just float; they kite surf on electricity.

I often think about those early explorers in the 1920s who first collected aeronautical spiders in traps tethered between biplane wings. Without the knowledge that we have today, I can only imagine how they made sense of finding spiders floating four kilometres in the air. Since then, spiders have been found even higher, and one type of spider has been found at an altitude of almost seven kilometres. And just to blow your mind a little bit more, when the spider was first discovered, it wasn't flying but walking.

On top of the Earth

The Himalayas are the highest mountain range on Earth. Their lofty peaks, reaching over eight kilometres in altitude, are some of the most inhospitable environments on the planet. Hundreds of people have died trying to reach Himalayan mountain summits. Even experienced mountaineers have fallen victim to avalanches, falls, or just the sheer physiological stress placed on their bodies when climbing in freezing temperatures at such high altitudes. At around 4000 metres, not even trees can survive and the landscape is covered in tough shrubs and grasses. The highest altitude at which humans can survive for long periods appears to be about 5000 metres. Above this, even shrubs and grasses can't survive, and plant life is mostly limited to crusts of lichen clinging to rocky surfaces. By 6000 metres, even lichens disappear and the mountains are seemingly lifeless. With no plant life to sustain a larger food web, few animals are ever found at these altitudes. Any animals that do manage to find their way up that high are usually temporary visitors, mostly migrating birds or intrepid humans hell-bent on conquering mountain peaks in the short period that their bodies can survive this high. One incredible exception to this is an animal that lives higher on mountain slopes than any other animal on the planet. Higher than mountain goats, higher than snow leopards, higher than most birds will ever fly lives (you guessed it) a spider.

Richard Hingston was the medical officer and naturalist on the British expedition to Mount Everest in 1924. This was the second attempt by the British to reach the summit. Ultimately, the mission was unsuccessful and the British would not reach the summit of Everest until Sir Edmund Hillary led his history-making expedition in 1953. Nevertheless, the 1924 expedition led to some surprising discoveries. When he wasn't busy looking after the health of the mountaineers and porters, Hingston was documenting the plants and animals he encountered along the way. In doing so, he collected thousands of specimens that were entirely new to western science. Amid all of these specimens, there was one that stood out. An animal that was found higher up the slopes of the Himalayas than any other creature – the Himalayan jumping spider.

Darting among ice-covered rocks, as high as 6750 metres in altitude, lives the Himalayan jumping spider, *Euophrys omnisuperstes*. At only half a centimetre in length, with slender legs covered in short, fuzzy hairs, this jumping spider doesn't seem like a particularly hardy creature to be found living above the clouds on the jagged, rocky slopes of the Himalayas. Hingston concluded that this new spider species was a permanent resident of mountain slopes, but he also considered that the spider could be there by accident. Knowing that spiders can balloon through the air, Hingston could have stumbled across some very unlucky spiders who were dumped into an inhospitable environment by a gust of wind. Since then, however, *Euophrys* has been found by other mountaineers and explorers on their way to Everest. So, it's unlikely that these spiders are simply accidental tourists and that *Euophrys* do survive on these icy peaks. No other animal has been found to permanently live at such high altitudes and *Euophrys* currently holds the world record for highest known permanent resident on Earth. Its scientific name, *omnisuperstes*, can be translated from Latin to mean 'highest of all'. There is one problem with this idea though. If only *Euophrys* live this high up the mountains, what do they eat? For populations of jumping spiders to persist at these altitudes, there must be other animals there for them to feed on. Again, this thought seemed to befuddle Hingston when he first encountered them and he made the imaginative suggestion that if jumping spiders were the only creatures there, they must survive by feeding on other jumping spiders. We can forgive Hingston for overlooking the mathematical impossibility of this suggestion. After all, low oxygen levels at high altitudes can make it hard to think straight. Back in 1924, no-one knew how animals could survive without a food web supported by photosynthesising plant life. It wasn't until the 1960s when ecologists such as Lawrence Swan, a prolific mountaineer and natural historian, would describe a new kind of food web that sustains life high on mountain summits, supported by the constant delivery of wind-blown debris, also known as aeroplankton.

When gusts of wind collide with the sides of mountain ranges, aeroplankton is strewn across the sides of the mountain, littering the slopes with small insects, pollen grains and other assorted organic matter. This deluge may provide the organic matter that can sustain small herbivorous insects and predatory spiders. It's likely that the hardcore Himalayan jumping spider manages to live in one of the most inhospitable environments on the planet by surviving on nature's aeronautical flotsam and jetsam. No-one has studied how the Himalayan jumping spider survives in this extreme environment. At these lofty altitudes, the number one priority of even the most enthusiastic explorers is their own survival and not understanding the minutiae of a jumping spider's daily routine. For now, the secret life of the Himalayan jumping spider remains a mystery.

Having conquered watery depths and the tallest mountain peaks, and traversed the skies on electric rockets, there is one more frontier that spiders have explored – outer space.

Chapter 9
SPIDERS IN SPACE

Skylab and the first spiders in space

When balls of fire began raining from the sky one morning in 1979, the small town of Esperance in Western Australia suddenly became world famous. At 2.37 am on 12 July, NASA's Skylab space station re-entered the lower atmosphere and fell to Earth over the south-west tip of Australia. A 40 kilogram hatch door landed just outside of Esperance and a two-metre-long steel and fibreglass cylinder crashed in a field near the small town of Rawlinna. A giant oxygen tank landed on a farm property so remote that it wasn't discovered until 14 years later.

When NASA first discovered that the now-abandoned space station was moving out of orbit, they needed to manage its fiery descent to Earth. It was impossible to predict exactly where it would land and NASA did their best to reassure people that they were incredibly unlikely to be hit by a piece of Skylab. Regardless, the idea of a space station the size of a house crashlanding on Earth sent people into a frenzy. Dick Smith, the Australian adventurer and business magnate, famously took out a $2.6 million insurance policy against a customer in one of his electronics stores being hit by a piece of Skylab. In the end, no-one was hit by falling space debris. As predicted, most of Skylab vaporised upon re-entry and most pieces that weren't destroyed fell into the ocean. The few chunks that made landfall were those that made Esperance famous. The descent of Skylab was world news because it was the first space station NASA had ever built, setting a precedent for future space stations like the International Space Station, which is still in operation. Skylab also set a world first when it was, for a short period, home to the first spiders in space. And they wouldn't be the last.

On 28 July 1973, two garden orb-web spiders (*Araneus diadematus*) named Arabella and Anita took off in a Saturn IB rocket from the John F. Kennedy Space Centre in Florida, with a little help from the three human astronauts on board. They were en route to the Skylab space station, where they would be observed in zero gravity. While this might sound like the plot of a bad sci-fi movie, there were sensible reasons for sending Arabella and Anita into space. To understand how life on Earth is shaped by ever-present forces like gravity, scientists can take gravity out of the equation by sending animals into space and studying how they behave in this strange environment.

To encourage meaningful connections between ordinary people and the important work that NASA does, NASA devised the Skylab Student Project; school students from across the United States were invited to design and propose experiments to be carried out on the space station. Of the 3409 proposals that were submitted, 19 were selected and approved for the Skylab missions. Judith Miles from Massachusetts proposed an experiment to study if spiders could spin webs in zero gravity. She worked with NASA scientists to design the experiment and custom-made enclosures where the spiders could spin webs and their behaviour could be filmed and photographed while on the space station.

Within a few days of the crew arriving at the Skylab space station, the web-building experiment began. Arabella was to be the first spider released into the open space of her experimental enclosure, and the astronauts on board bore witness to a spider experiencing the weightlessness of space for the first time in history. At first, the spiders made 'erratic swimming motions' through the air before taking hold of a solid surface but it didn't take long for them to show signs of adapting to their new environment. Within a day Arabella had built a small web in the corner of her enclosure. The next day she had built a complete web that spanned the whole enclosure. The more webs the spiders built, the more normal they looked. Soon the spiders spent less time swimming in zero gravity and more time crawling along web threads and hard surfaces, apparently adapting to their new conditions. The webs they spun never quite seemed to match the kinds that they build on Earth. They were limited in size by their enclosures, and the webs tended to have thinner silk and irregularly spaced threads.

As remarkable as this story is, we need to be honest here and admit that this wasn't the greatest piece of spider research ever conducted. While I don't doubt that NASA's scientists and astronauts are incredible engineers, mathematicians and physicists, arachnologists they are not, and animal husbandry was simply not one of their strengths. They only brought two small flies along as food for the spiders. Despite offering the spiders small chunks of meat from their food provisions, the spiders showed no interest. Sadly, Arabella and Anita didn't survive long enough to make the journey back to Earth. Their bodies were brought back and given to the Smithsonian's National Air and Space Museum in Washington. The experimental protocols weren't exactly the greatest either. The photographs that the astronauts took of the spiders' webs didn't have any scale reference in them, and the cramped conditions inside the space station meant that the cameras were placed too close to the enclosures; the photos didn't capture the whole webs, so we have no idea how big they were. Photos of Arabella and Anita's webs taken before the mission were of such poor quality that they couldn't be used for comparative analysis.

To be fair, I shouldn't nitpick too much; these were the first spiders in space, after all, so the whole experiment was going to be a learning experience. What was achieved, in the end, was very promising and provided the foundations for future spider space missions. The next one was going to be bigger and better, with more spiders, more experiments, and better technology. It was going to be amazing. It was going to end in tragedy.

The space shuttle mission

It took another 30 years before the next crew of spiders would find their way into space. In 2003 eight Australian garden orb-web spiders (*Hortophora transmarina*) launched into space, this time not on a rocket but a space shuttle. They embarked on a 16-day mission along with seven human astronauts. The spider habitats were new and improved, and the astronauts had the ability to release one spider at a time into the experimental arena. These enclosures came with automated infrared night lighting, temperature sensors, humidity control, dedicated cameras that took images throughout the entire journey and, most importantly, extra flies to keep the spiders alive through the entire journey.

With 30 years of improvements in space technology to help them, the astronauts could more robustly replicate the experiments conducted in 1973, and analyse if and how spiders could spin webs in the absence of gravity. This mission involved a multi-institutional and international collaboration between NASA, Spacehab, BioServe, RMIT University, the Royal Melbourne Zoo, and a team of enthusiastic students from Glen Waverly Secondary College in Melbourne, Australia. Just like the first attempt at studying spiders in space, this mission would engage school students in real-life space exploration. What started as a one-year project grew into a three-year project when the shuttle launch was delayed multiple times. Over these three years, the students had an unsurpassed opportunity to be at the heart of an international space exploration mission. This culminated with the students being flown to the United States to load their experimental apparatus into the shuttle payload, and watch the launch of the shuttle from the Kennedy Space Centre. The plan, at the time, was for the spiders to return to Earth, and be flown back to Melbourne where they would go into quarantine and be put on public display as the first spiders to have survived a mission to space.

Overall, the mission was a resounding success; the astronauts completed all their experiments, including observing the spiders building webs in space. Low-resolution images of the webs were broadcast back to Earth, and the astronauts collected silk samples and high-resolution photos for further analysis. With the mission completed, the astronauts readied the shuttle for landing and the student scientists, now back in

Australia, waited eagerly for the return of their data and their beloved spidernauts. Sadly, tragedy struck and the astronauts, and spiders, would never return.

You may have noticed that I haven't mentioned the name of the space shuttle these spiders travelled on; it was called *Columbia*. It's famous for being the first space shuttle ever launched and, 28 missions later, for its tragic end in one of the greatest disasters of space exploration history. After completing its twenty-eighth mission, *Columbia* was due to land on 1 February 2003. All was going as planned until, about 15 minutes away from its projected landing on Earth, NASA suddenly lost radio contact with the shuttle.

In southern Texas, keen observers were watching the skies, hoping to catch a glimpse of the space shuttle re-entering Earth's atmosphere on its way to Florida. When *Columbia* finally came into view, all seemed to be going to plan; a bright white vapour trail drew a slow, straight line across clear blue skies. Soon though, this white line was punctuated by what looked like white flashes, and the straight white line began to split into a cluster of white threads. The white vapour trail became clouds of smoke and it was clear that the shuttle was returning to Earth in pieces.

NASA eventually discovered what caused the tragedy. During the launch a large piece of debris collided with the left wing of the shuttle. This caused a small fracture in the heat protective shields on the wing that didn't cause any issues until the shuttle re-entered the Earth's atmosphere. These heat-resistant shields would normally have protected the shuttle from the extreme heat caused by the friction between the atmosphere and the descending shuttle. The small fracture led to super-heated gases flooding into the shuttle interior, disintegrating the craft and killing all seven astronauts on board.

A search and recovery mission collected all the remains that could be found of *Columbia*. None of the data or equipment from the student research projects could be recovered. From the data sent back during the mission, the Australian students could conclude that their spiders successfully built webs in microgravity. The rest of their data, silk samples, high resolution images and spiders were lost.

When *Columbia* took to the skies it heralded a new era of space exploration. Its tragic end reminded us that exploration can sometimes come with great risk and great sacrifice. The year 2023 marked the twentieth anniversary of the *Columbia* disaster. It was commemorated with sombre ceremonies across the globe that acknowledged the great sacrifice made by the seven astronauts who died on board. And in a park in the suburbs of Melbourne, a small group of old school mates met up for a reunion with a few of their teachers to reminisce about an incredible experience they had 20 years earlier. The excitement of sending spiders into space, followed by the trauma of watching the *Columbia* tragedy unfold, has indelibly bonded this group of students and teachers. Even after 20 years they have never forgotten each other and have never forgotten their dear space-spiders.

Spiders on the International Space Station

In 2008, two more orb-web spiders launched into space on the *Endeavour* shuttle. One *Metepeira labyrinthea* and one *Larinioides patagiatus* (affectionately named Elmo and

Spiderman) were on their way to the International Space Station (ISS). On arrival, they were transferred from the shuttle to the space station and very quickly made news headlines when the crew noticed that one of them was missing. *The Times* in London ran the dramatic headline 'NASA loses spider on International Space Station', sending imaginations running wild. In reality, the situation wasn't quite as exciting as it sounded. The spiders were brought into space in a specially made payload that had multiple chambers inside it. The missing spider had simply managed to find its way out of its initial holding chamber and into the larger experimental chamber. It was a head-scratcher for the astronauts on board and the company that built the enclosures, but not the nail-biting escape story readers of the paper were expecting. Two days later, and just in time to help the crew celebrate the tenth anniversary of the ISS, the two spiders had built a pair of organised, symmetrical-looking webs. Just like their predecessors, these two spiders adapted remarkably well to weightlessness within a matter of days.

As with the previous missions, this experiment was run as an educational outreach project. School groups across the United States could tune in to updates on the spiders' web-building and compare them to the webs of spiders they were keeping in their classrooms. Unfortunately, the observations had to stop after only eight days due to an unexpected issue with space slime. While all of the attention was on the two spiders, there were some other animals on board that were being overlooked – the fruit flies brought along as live spider food. Having learnt from earlier missions, the enclosures contained a large number of fruit flies to keep the spiders well fed. There was a small agar plate containing fly larvae that would develop into adults over time, creating an ongoing food source for the spiders. Apparently, the fruit flies were also capable of adapting to space travel and started breeding inside the spider enclosure. Pretty soon the clear walls of the enclosure were covered in slimy mucus left behind by fly maggots roaming around inside.

There was more in store for the spiders, and the slimy lining turned out to be a silver lining. Since the fruit flies were breeding inside the spider enclosure, their food supply could extend longer than originally expected. The spiders were initially scheduled to return to Earth after two weeks, when the *Endeavour* shuttle went home. Instead, they stayed on board the ISS. What the spiders did during their time in space remains a mystery as the maggots and fly carcasses in the enclosure made direct observations difficult. Three months after the spiders arrived on the ISS, NASA reported that a camera inside the enclosure had detected the movement of a live spider on its web. How much longer they survived isn't known; both spiders were deceased when they arrived back on Earth over a month later with the space shuttle *Discovery*.

What started as a short experiment on web-building behaviour in space turned into a pilot study showing that spiders can live in space for months at a time. In the end, the spiders' stay in space was longer than that of many professional astronauts.

Bringing space spiders to the whole world

The shuttle *Endeavour*, which brought Elmo and Spiderman to the ISS in 2008, would soon carry two more spiders into space. In 2011, Gladys and Esmerelda, two *Trichonephila clavipes* spiders, were launched into space and transferred to the International Space Station, where they would be observed over the next 60 days.

Like before, this experiment was an educational initiative, but instead of just sharing the research with a small group of schools, this time the adventures of the space spiders would be shared with the whole world. Photos of the spiders building webs in space were downlinked to Earth as the experiment was being conducted and were made freely available to view online so students and teachers from across the globe could get ongoing updates on how the spiders were faring in space.

After the false starts and missteps of previous trials, this fourth attempt at studying spiders in space would be the most successful so far. The experiment was planned to last 60 days, much longer than the previous eight days of observations. This time, the breeding fruit flies were kept in a separate compartment so that adult flies could be gradually released into the spider enclosures, and maggots couldn't slime over the observation windows. On Earth, *T. clavipes* webs are elongated with a central hub that sits much closer to the top of the web than the bottom, and the spiders sit on the hub facing directly towards the ground. In space, the webs were more symmetrical, with a central hub in the middle of a mostly round web, indicating that gravity influences the larger structure of webs on Earth. The spiders in space couldn't position themselves facing downwards on the web, since in microgravity there isn't really an up and a down; instead they tended to just point themselves away from the enclosures' light source.

Other major milestones included a new record set by Esmerelda for the most spider webs built in space by a single spider (34 webs) and a record set by Gladys for the longest time spent alive in space by a spider (65 days). The mission was so successful that Gladys (who actually turned out to be a male spider) survived the trip home to Earth. And just to make the moment even more special, Gladys returned home on the very last space shuttle landing. On 21 July 2011, the space shuttle *Atlantis* touched down at the Kennedy Space Centre, bringing with it the first spider to have ever returned home alive from space (and the less fortunate Esmerelda of course), and marking the end of 30 years of space shuttle exploration.

Jumping spiders in space

As was established at the beginning of this book, glossy-eyed miniature jumping spiders are irresistibly adorable. So, when video sharing website YouTube and computer manufacturer Lenovo went looking for ideas for scientific experiments to be performed on the ISS and live-streamed online, how could they resist sending jumping spiders into orbit? Of around 2000 experimental proposals received from teenagers across the globe only two winning entries were selected: one that proposed investigating bacterial growth in space, and another sent in by Amr Mohamed from Egypt that proposed investigating if and how jumping spiders catch prey in a weightless environment.

In July 2012, one red-backed jumping spider (*Phidippus johnsoni*) named Nefertiti and one zebra jumping spider (*Salticus scenicus*) named Cleopatra launched on board the uncrewed *Kounotori 3* rocket from Tanegashima Space Centre in Japan. They arrived at the ISS six days later and would spend the next 100 days orbiting the Earth. Whereas previous missions observed the web-building behaviour of spiders, this mission was focused on whether jumping spiders could still pounce on their prey in microgravity. Two months into the mission, viewers from across the globe tuned into a special livestream event hosted by science celebrity Bill Nye to find out the results of the two experiments. During the livestream, astronaut Sunita Williams called in from the ISS to show the spider Nefertiti happily wandering about her enclosure, and confirmed that the spiders had adapted to microgravity and were successfully stalking and pouncing on their prey.

By the end of the mission, the two spiders had set new records, having spent 100 days in space on board the ISS, more than any other spider before them. And when the two spiders finally returned to Earth, they made history by returning on the first commercial craft to supply the ISS: the SpaceX *Dragon*. While both spiders survived the entire length of the mission, Cleopatra died shortly after landing. Nefertiti however was in good condition despite taking a little while to re-adapt to Earth's gravity. A month after her return to Earth, Nefertiti was given a special enclosure and went on display at the National Museum of Natural History in Washington so that people could meet a real-life spider-naut. After 100 days in space, Nefertiti's enclosure in the museum should have made for a long, cosy retirement. She died four days later.

So, after five space missions and sending 16 spiders into space, what have we learnt from all of this? It depends on how you look at it really. An arachnologist would likely be unimpressed by the datasets, as they haven't contributed all that much to our understanding of spider biology. The scientists involved in these missions are well aware that they aren't conducting the most rigorous arachnological studies. That was never the point. Spacehab and BioServe, the two companies responsible for coordinating and engineering a number of these missions, have stated very clearly that the programs were 'not designed to provide world-class research as the final outcome'. Any knowledge gained was secondary to their main goal, which was teaching people about space and engaging them with real-world space exploration.

I argue that there are two take-home messages from these space spider stories. Firstly, that spiders can adapt so quickly to weightlessness and the completely alien environment of a space station is testament to their flexibility and tenacity. If you ever had any doubts that spiders were anything more than simple robotic creatures, then these adaptable eight-legged astronauts must put those doubts to rest. Secondly, as I have argued throughout this book, spiders are awesome. So awesome that they are used for promotion and engagement by NASA and by YouTube, which is owned by Google, the biggest advertising agency to have ever existed. Neither NASA nor YouTube has been swayed by inaccurate spider stereotypes. They knew enough to realise that people's natural fascination with these incredible creatures makes them the perfect vehicle to promote space exploration.

If we actually want to learn more about spiders through space exploration then we need to address the lack of arachnologists in space. Perhaps next time we send spiders into space, we can also send some animal behaviour experts. If anyone at NASA is reading this, please accept this as my expression of interest for upcoming missions. I look rather fetching in a jumpsuit, and what I may lack in aeronautics experience I more than make up for with my can-do attitude. I look forward to hearing from you.

Chapter 10

PHOBIAS AND FABLES

Spring in Australia is celebrated with a sacred household ritual – the huntsman-removal ceremony. Upon whence a spider is sighted, a call is made throughout the house: 'A spider!' the witness doth call. Upon hearing the call, the 'Catcher' (a designated, often senior, family member) will respond; 'Get me a container! And a piece of paper!' At this call, the family descends upon the kitchen and returns to the Catcher with an assortment of cups, glasses and old plastic containers. The Catcher inspects the offerings and carefully selects an appropriate receptacle befitting the size of the visiting spider.

The family will gather to watch as, slowly, the Catcher places the receptacle over the waiting spider. A suitable sheet of paper, perhaps an old electricity bill or envelope, is delicately slid between the wall and the receptacle. If done correctly, the spider will acquiesce and step onto the approaching paper platform. The Catcher will then place their hand under the piece of paper and, with great dexterity, lift the spider from the wall. With the spider now tenderly encased betwixt the paper and the receptacle, the Catcher will present the spider to the family for inspection. The family may remark in awe on its size and hairiness. Where appropriate, the Catcher may invoke a humorous response by pretending to drop the spider, or perhaps liken the spider's face to an unfortunate relative.

The Catcher then takes the spider outside to select a new home for their houseguest – perhaps a lush flowerbed, or the patch of grass behind the bins. With a skilful flick, the paper is drawn away from the receptacle and the spider is released safely into the garden. The ritual is then complete, the spider is left to enjoy its new home and the family may go about their business as they await the arrival of their next huntsman-guest.

Quite a few years ago, I was visiting my niece who, like many eight-year-old girls, hadn't yet fully grasped concepts such as gravitational pull and grievous bodily harm. She was the sort of kid wise enough to know that the best way to learn how to rollerskate is fast and down steep hills. Not one to stifle such ambition, I made sure to be the uncle who would happily adjust her helmet, tie her laces, and wish her good luck. After eight years of successful uncle-ship, and no major limbs lost, I was left surprised when one day I witnessed her reaction to a small huntsman inside the house.

A spider was found and, as per tradition, the call was made and I began the huntsman removal ritual. With a plastic container in one hand and a piece of paper in the other, I gazed upon our humble guest. She was a beauty. Perched motionless at around chest height on the dining room wall, her fangs glistened in the warm afternoon sun. I looked deep into her small black eyes and, I think, she looked back into mine. Then a thought occurred to me. I turned to face my family where my niece, the future of our clan, watched on. I knew then it was time for me to pass the Tupperware on to the next generation.

'Wanna catch a spider?' I said and handed her the container. Sensing the gravity of the moment, her face beamed. Eagerly she took the container and walked towards the wall. I watched on proudly as she stepped closer and closer, and then, stopped. Suddenly, her smile was gone and she looked concerned. 'Actually, I've changed my mind', she said. And walked away. I didn't understand. Why did this explosively confident young child suddenly decide that this harmless spider was too much to handle?

It turns out that I am not the only one perplexed by spider fears. For psychologists who study the causes and treatments of fears and phobias, spiders are puzzling. In many ways, spider fears are similar to other fears; they are more common in children than adults, and more likely to be experienced by women than men. What makes them special is how pervasive and inexplicable they are. Phobias often rank as the most common mental disorders globally. The most common types are called 'specific phobias': these are phobias that have an identifiable trigger (e.g. spiders, heights, flying), as opposed to more generalised sources (e.g. social phobias, fear of leaving the house). Of all the known specific phobias that people suffer from, animal-centred phobias are the most common, and out of these spider phobias top the list. Depending on the survey, snakes usually come in at a close second. These same patterns seem to be true for clinically diagnosed phobias, and when talking about people's general fears and dislikes that aren't significant enough to be considered phobias. Given how common spider fears are, they have been the subject of much research, but it almost seems like the more they are studied the less they make sense.

One traditionally touted explanation for a fear of spiders

is that it's a protective response, shaped by evolution. The logic is that since spiders are venomous and some are potentially harmful to us, our ancestors who were genetically programmed to fear spiders were more likely to survive and pass on their spider-phobic genes. It's an idea that seems intuitive but just doesn't hold water. As we saw a few chapters back, spiders are, on the whole, harmless. So benign that it takes a huge stretch of the imagination to think that they could shape the evolutionary trajectory of our brains. What about the few potentially harmful species, like the Australian funnel-web spider or the South American wandering spider? If our evolutionary history with these spiders is enough to have led to justifiable spider fears then we should see this reflected in geographic and cultural patterns of spider fears. Again, there is no compelling data to show that spider fears across cultures vary predictably with the types of spiders those cultures would have come into contact with, though data on this topic is pretty sparse, as good cross-cultural datasets are extremely hard to collect.

Even if we assume that the extremely low chances of a severe spider bite were somehow enough to shape patterns in the evolution of the human brain, then we would expect to see similar patterns in our fears of other animals that are just as, or even more harmful to us. For example, other venomous arthropods – like ants, bees, wasps and scorpions – can pose just as great a threat to us as spiders – arguably more given that things like bee venom have evolved to be used for defence against large predatory animals. Yet studies that surveyed people's fear responses to these animals showed that they pale in comparison to how terrified people are of spiders. Even in surveys that have studied people's fear responses to large predators like bears, lions and sharks, spiders still come out on top as the most fear-inducing animal. As I said before, it just doesn't add up.

When psychologists study animal dislikes and phobias, they identify two separate yet related emotional responses: fear and disgust. *Fear* responses are experienced when a stimulus presents a threat of physical harm or death. For animal fears, this could be in response to a large, predatory animal. Whereas *disgust* should be felt in response to something associated with disease or illness. Animals like rats, mice and maggots often rank as highly 'disgusting', presumably because of their potential to carry diseases or their association with decomposition and decay. Generally, something that triggers a fear response in a person doesn't also trigger a disgust response; people who are afraid of heights generally don't describe them as disgusting. Again, spider fears are peculiar because people tend to score them high on both fear and disgust. A fear response is hard to justify, given how harmful they really are. Similarly, a disgust response is hard to justify given that there is no clear connection between spiders and sickness. They aren't vectors for any human diseases, and they are not associated with decomposition or decay. Overall, they seem to be pretty tidy little creatures and yet they are ranked as more disgusting than disease vectors like mosquitos and rats, or animals that associate with decomposition and decay like flies, cockroaches and beetles.

So, what can be done about spiders and our unfortunate relationship with them? It depends on whether we are talking about society's general dislike of them or about an individual's clinically significant phobia. Either way, there is hope.

Dealing with a spider phobia

If you have a genuine phobia of spiders and you are reading this book, then congratulations, you should be very proud of yourself. Though chances are that if you are reading this, you don't have a spider phobia. You may be a bit uncomfortable with them, or just find them somewhat creepy, but a genuine phobia can be a crippling condition. Diagnosis of a phobia is a job for a trained psychologist, but in general terms, they are fears so extreme they prohibit you from going about your daily life. Seeing a spider in the office might make one person feel a bit spooked, but it might cause another person to be unable to ever go back to work for fear of seeing spiders; that's when we're deep in phobia territory.

As much as I have been making light of spiders in this book, genuine phobias are no joke. They can have severe consequences for a person's wellbeing, including their social lives, mental health and socioeconomic status. In 2012, the healthcare costs just for specific phobias across Europe were estimated to be almost €20 billion. Furthermore, phobias can be associated with other mental health conditions including depression, anxiety and even substance abuse. Often people can recognise that their fears or reactions are irrational, and the lack of control they have over their irrational responses is an additional source of anxiety. In many cases, the most direct way to deal with a phobia is to simply avoid the stimulus. This is relatively easy for someone with a fear of flying, for example. They can live relatively unaffected by their condition by simply never stepping foot on a plane. This degree of control over their phobia isn't possible for people suffering from spider phobias. Since spiders are so common and so abundant in our homes and gardens, simply avoiding the trigger isn't an option, and living with an extreme spider phobia can mean living in a state of constant fear. For this reason, professional intervention is often necessary.

The standard treatment for any phobia is exposure therapy. As its name suggests, it involves a person safely and willingly exposing themselves to the subject of their phobia to desensitise themselves. Over time and multiple treatment sessions the intensity of the exposure increases. Depending on how afraid someone is of spiders, this could begin with something as simple as talking about spiders or looking at a picture of a spider from a distance. Even these situations can be such a barrier to people that there has been research into 'spiderless' exposure therapy, where patients can begin by looking at pictures of things that look vaguely spidery, like a chair with long legs. Under the guidance of a trained professional, the treatment can progress towards having the person be able to look at a live spider in an enclosure from across a room then moving the spider gradually closer until the person no longer perceives it as a threat. You might hear stories of psychologists boasting that exposure therapy can quickly take someone from being unable to even look at a spider, to being able to have a live tarantula crawl along their arm. As impressive as this sounds, this needn't be the outcome of everyone's treatment. Unless you have career aspirations of being an arachnologist or plan on milking spider silk for profit, then you shouldn't ever need to let spiders crawl all over you.*

* If you are in a position where you have to handle a spider, you can use the traditional glass container and piece of paper technique. Whatever you do, be kind, be appreciative, and love spiders for the majestic beasts that they are.

Inevitably, undergoing exposure therapy requires people to face their fears, and experience some level of discomfort. Many people suffering from phobias will never seek help for this very reason, and of those who do initiate treatment many quickly drop out. New technologies are being used to help make the treatment of spider phobia more accessible and affordable. Virtual reality techniques are being used to treat many different phobias without the patient ever actually facing the source of their fears. By donning a VR headset, a person can enter a virtual reality environment where they and their therapist have exacting control over the size, shape, number and behaviour of the virtual spiders they are exposed to. You could begin by entering a virtual reality kitchen with a small motionless virtual reality spider on the wall, and work your way up to running around a virtual reality haunted house, with monster-sized spiders that chase you. Again, this probably isn't necessary for most people but I imagine it would be a good way to bulletproof your new relationship with spiders.

If, like most people, you don't have easy access to virtual reality headsets and computers, there are several smartphone apps that can let you interact with virtual spiders in your own home and at your own pace. These apps are a useful option for people who might not be able to access a trained therapist, or who might still be struggling with some stigma around seeking help from a professional. As with all smartphone apps, they're of varying quality and most of them haven't been backed by research. Unlike virtual reality apps that present you with a virtual environment, augmented reality apps use your device's camera to show you images of the room around you with virtual spiders superimposed on different surfaces. I tried out an augmented reality app called Phobys, developed by the

University of Basel and supported by clinical research. Using the app on my phone, I could view the tabletop in front of me and, when I was ready, a small brown virtual spider appeared on my table. I could move in to see it closer and walk around the table to view it from different angles. The app progresses you through different levels of spider interaction right up to prompting you to 'touch' the virtual spiders. I can easily see how these kinds of apps could be a great option for someone taking their first steps toward becoming more comfortable with spiders in the safety of their own home, without having to face a real spider, and at a cost that is a few orders of magnitude cheaper than some sessions with a psychologist.

At the end of the day, the biggest predictor of treatment success is the attitude of the individual: their willingness to undergo treatment in the first place, and to continue in the face of the inevitable distress it will cause. While we don't know exactly what causes spider phobias, or why they are so common, we know a lot about how to treat them and can do so very effectively. The other issue at hand – society's general dislike of spiders – is a little trickier to deal with, and might take an enormous group effort to tackle.

Spiders and society: Will we ever get along?

The jury is still out on whether spider fears are a feature of our biology. Another idea that has been put forward by psychologists suggests society's disdain for spiders is simply an unfortunate symptom of some very infectious ideas. In the early 1990s, psychologist Graham Davey was trying to understand why spiders would be perceived as 'disgusting' by people who were afraid of them, when other even more dangerous animals weren't. Apparently, in Europe in the Middle Ages and throughout the time of the bubonic plague, there were myths circulating that spiders could spread disease and contamination. It was said that if a spider had come into contact with any food or water then it must be discarded for fear of contamination. The actual cause of the plague (a bacteria carried by fleas living on black rats) wasn't known for centuries, so there was plenty of time for different myths to spread. Spiders were just one of these misguided ideas along with other potential causes such as foreigners, too much exercise, excessive bathing, improper alignment of the planets, and olive oil.

Despite there being no link between spiders and disease, the idea was planted and, Davey suggests, has been passed down through generations and persists to this day. We know that there is a heritable component to animal fears. A person who fears spiders is likely to have parents who fear spiders. It's common for people to point towards their parents' behaviour as a source of their anxiety and surprisingly uncommon for adults suffering from a phobia to be able to recall a traumatic incident that may have triggered their fears. The prevalence of spider fears in women, and the fact that women tend to rank spiders as more 'disgusting' than fearful compared to men, has led to speculation that spider fears are inherited by mimicking maternal behaviour. Though again, the data is limited. If this is true then the whole idea of spiders being scary is just a particularly infectious idea. Spider fears might just be a meme. They may simply be ideas that, like all good memes, spread because they spread because they spread.

It's not hard to imagine spider fears being a dangerous self-perpetuating meme. One can easily point to villainous portrayals of spiders in popular culture, such as Shelob from *The Lord of the Rings*, Aragog from *Harry Potter*, or the menacing hordes of spiders in cult classic science fiction films like *Arachnophobia* and *Eight Legged Freaks*. *The Black Spider* by Jeremias Gotthelf is a classic book from the 1800s that is often credited as being one of the earliest horror novels; spiders sent by the devil plague a small town and can kill at a touch. Flights of fancy can seem easy to ignore but these fictional narratives, unfortunately, plague other aspects of culture. I can't help but cringe when media outlets rely on clichéd spider stereotypes to sell their stories. Even articles as benign as a story about a new species being discovered will grab readers with a headline about scary monsters or an obligatory sentence describing the spider 'like something from *Arachnophobia*'. Sadly, I have even seen these headlines used by popular science outlets, and nature and wildlife magazines, which have fallen into the trap of using clickbait in favour of accurate reporting.

This might sound like an overreaction but we know from experience that popular culture can have huge impacts on how we perceive animals. Unfortunately, the fictional stories we tell about animals can take precedence over facts in our minds and lead to devastating outcomes. The most famous example of this is a phenomenon called the '*Jaws* effect'; and you can probably guess why.

The '*Jaws* effect' and why storytelling matters

The movie *Jaws* was released in 1975 and became an instant hit that changed cinema, and the fates of sharks, forever. Based on the 1974 novel of the same name by Peter Benchley, it tells a fictional story about a 'rogue shark' that spends its time swimming around a small-town beach, hunting down humans. It's a fantastical thriller based on ideas so outlandish that it should have never been taken seriously. But take it seriously people did. The tantalising idea of a menacing shark with a thirst for human blood took hold in society. It instantly changed how we think about sharks, and the film – and its blockbuster success – has been accused of contributing to declining shark numbers worldwide.

The 'rogue shark' in *Jaws* is a larger-than-life movie monster. There is no evidence that sharks intentionally hunt humans, there is no evidence that sharks will maintain a territory where they continually hunt until they exhaust all the food in that area, and there is absolutely no evidence that hunting down an individual shark will protect other people from future shark bites. Sadly, these are the ideas about sharks that continue to be perpetuated in popular media. Whenever a shark attack occurs it immediately triggers a media frenzy of articles that repeat the same old stereotypes. Phrases along the lines of 'like a scene from *Jaws*' are repeated ad nauseam, and stock photo images of open-mouthed great white sharks are plastered across front pages and social media feeds.

The false idea of a rogue shark actually stems from an Australian surgeon who had the idea that a single shark could be responsible for a number of attacks in the 1950s. This surgeon was speaking way out of his area of expertise. Just as we have already learnt that we shouldn't trust doctors' advice on spider behaviour, we similarly shouldn't pay too much attention to their thoughts on shark behaviour. In reality, many shark species are listed as endangered or vulnerable by the International Union for the Conservation of Nature (IUCN); their global population numbers have been in sharp decline. The greatest threat to sharks is humans. Worldwide they are being fished for food and sport faster than their populations can recover, leading to many species facing potential extinction. Sadly, the idea that sharks are a threat to humans, as told in stories like *Jaws*, is a much more infectious idea that continues to infiltrate our conversations around sharks, despite it being pure fantasy. Immediately after the film's release, there was a surge in the popularity of fishing for sharks across the United States, with sports fishers boasting their abilities to hunt down bloodthirsty human-killers that were bigger than their boats. Why? Because that's what the heroes from *Jaws* did, and they looked cool doing it.

Even more horrendous is the fact that the idea of hunting down 'rogue sharks' has been implemented in official government policies. Dr Christopher Neff originally coined the term '*Jaws* effect' when investigating the strange decisions of the Western Australian government in response to a series of recent, and highly publicised, shark attacks. In an article published in the *Australian Journal of Political Science*, Neff outlined how public pressure and ignoring evidence led to official government policy that was apparently based on rogue shark myths rather than any scientific evidence. Despite expert opinions that it would be near impossible to positively identify a shark that was involved in a previous attack, and that the sharks would be highly unlikely to stay in the same area, the government issued a 'special order' to attempt to find, identify and kill the sharks responsible. Just like they did in the movie *Jaws*. Despite the lack of scientific evidence endorsing it, it became enshrined in formal policy and led to public resources being spent on hunting down non-existent rogue sharks in the interests of public safety. Peter Benchley, the author of *Jaws*, even spoke out against the actions of the Australian government, and spent much of his later career working to promote ocean conservation, including writing a book called *Shark Trouble*, a memoir acknowledging the mistakes made by *Jaws*, and decrying the damage that can be done by sensationalist media.

It doesn't take a huge stretch of the imagination to see how spiders might be suffering the same fate as sharks. Just like the '*Jaws* effect' puts fiction over fact when talking about sharks, the fantastical idea that spiders are scary and deadly creatures has taken hold in our collective consciousness, and takes precedence in our minds over the much more nuanced truth. When I talk to people about spiders, and I hear the stereotypical responses of 'Gross!', 'Eww!' or 'Nope!', I have often wondered whether these are genuine responses, or whether people are simply responding in a way they *think* they are supposed to. Maybe, in the absence of better spider anecdotes to tell, some people immediately default to a socially acceptable behavioural shorthand. You may be thinking, *who cares? It's all a bit of fun; who doesn't like playing around with a creepy-crawly spider story?* Try telling that to someone with a crippling spider phobia. If the perpetuating whirlwind of negative spider stories is maintaining unfounded anxieties throughout society that can snowball into serious mental health conditions, then spider myths for some are a much bigger problem than we want to believe. Personally, I get frustrated at the injustice of it all. I find it such a shame that these fascinating, harmless and often beautiful creatures are the focus of so much fear and misrepresentation. The possibility that this could be boiled down to some simple misunderstandings and tenacious myths feels like a wrong that needs to be righted.

If spider fears are rooted in our biology, then that's a sticky problem that's difficult to treat. If, however, spider fears are a cultural problem, then there is a silver lining – we can do something about it. As the *Jaws* films have shown, scary animal stories can have unintended consequences, but as the discovery of humpback whale songs in the 1970s proved, the right stories can change how we feel about animals for good. Better spider stories could make all the difference for us *and* spiders.

Let's tell better spider stories

Throughout history, spiders have been portrayed in a much more positive light than we may think: as beacons of tenacity, creativity and cleverness. A Scottish legend tells of King Robert the Bruce who, after being defeated at the Battle of Methven in 1306, was forced into hiding and took refuge inside a cave for months. While in the cave, he watched a spider struggling to build its web. Each time the spider fell from the cave walls, it persisted and tried again until its web was finally complete. From the teachings of this humble yet tenacious creature, the king was inspired to gather his troops and return to battle against the invading English. King Robert went on to claim victory and independence for Scotland in the Battle of Bannockburn in 1314. He continued to reign until his death in 1329 and is still remembered as a national hero. No-one knows what happened to the spider, but we remember it too.

An Islamic legend tells a similar story about the prophet Mohammed hiding from soldiers in a cave. A spider wove a thick web, covering the cave entrance. When the soldiers came near, they decided that no-one could be in the cave as they would have had to break the spider web to get in.

There is a Ukrainian tradition of having decorative spider ornaments on Christmas trees that stems from a fable about a poor family who had nothing to decorate their *yalynka* (tree). Taking pity on them, a house spider (or spiders) covers the tree in silk in the middle of the night when the family is asleep. In the morning the webs have transformed into glistening silver. It is believed that this is where the tradition of having tinsel on Christmas trees comes from.

Ancient mythology recognised the skilful weaving of spiders and tales of clever spider characters have made their way into several well-known tales. The Roman poet Ovid told a tale of a mortal woman Arachne who was such a skilled weaver she was challenged to a contest with Athena, the goddess of handicraft, wisdom and warfare. In one version of this tale Athena loses the battle and, in a rage, tears Arachne's handiwork to pieces before transforming her into a spider, where she would weave forever. The class of animals to which spiders belong, Arachnida, derives its name from this ancient myth. The Spider Woman is an important deity for the Diné – the Navajo people. Her story is one of creation: she was a skilful weaver who wove together the fabric of the universe on a loom built by her husband Spider Man. Her weaving skill was passed on to the Diné and she is revered for gifting people with the skills and knowledge to prosper.

The D'harawal people of south-eastern Australia tell a dreaming story of two spider siblings, Mararan and Marareen. Mararan was fat and lazy and lived in a cave while Marareen spent her time in the trees busily weaving. In her book *D'harawal Dreaming*, D'harawal elder Frances Bodkin says this story was told to teach people the difference between potentially dangerous ground-dwelling spiders, like mouse spiders and funnel-webs, and harmless tree-dwelling spiders like orb-web spiders.

We needn't look back so far in history to see how positive spider stories weave their way into art and culture. From an early age, we are taught to empathise with the hard-working Incy Wincy spider who keeps climbing up the waterspout only to get washed back down again and again. One of the most famous modern spider tales is E.B. White's *Charlotte's Web* where Charlotte, a cunning barn spider, hatches a plan to save Wilbur the pig from being slaughtered by weaving words of praise about him into her web. These miraculous webs soon turn Wilbur into a local celebrity and his life is spared. And of course, we can't forget the multitudes of *Spider-Man* comic books, shows and films with their cavalcade of spider-themed heroes, including Spider-Man, Spider-Woman, Spider-Girl, Spider-Boy, Spider-Gwen, Spider-Ham (yes, it's a pig), the list goes on! Spider-Man first appeared in comic books in 1962, and since then his popularity has soared, taking the superhero, and the spin-off characters, from the pages of comic books and into video games, toy lines, cartoons, and even a short-lived Broadway musical. The popularity and cultural impact of the Spider-Man films is hard to ignore. The 2002 Spider-Man film starring Tobey Maguire and directed by Sam Raimi was such a critical and box-office success that it is often given credit for redefining superhero films, and paving the way for the superhero cinematic universes that dominate the box office today. A long string of blockbuster Spider-Man films followed and there are more in the making. This isn't just good news for comic book fans, but it might also be good news for arachnophobes. In 2019 a team of researchers from Italy showed that people's spider-phobic symptoms were reduced after watching clips from Spider-Man movies. For the study, they recruited a cohort of people and surveyed their fear responses to both spiders and ants. They then showed the participants short clips from either the 2002 Spider-Man film or the 2015 Ant-Man film, and then surveyed their fear responses again. In comparison to control groups, the people who watched clips of the ant- and spider-themed films showed a noticeable reduction in their fear responses. The researchers argued that these fantasy films could portray insects and spiders in a positive broader context, leading to more positive associations between the audience and the sources of their phobias.

Since positive spider stories have the power to change how we feel about them, wouldn't it be great if someone wrote a book full of cool spider stories? We could see if it made a difference in how people feel about them. Wait a minute! You're reading it! Remember the Spider Vibes Survey from the beginning of the book? What was your score? Go back and revisit the questionnaire now that you are almost at the end of this book. Is your score the same or has it changed? Let me know what you think. I'd love to know if this collection of spider stories has made a difference, even if just a little.

Chapter 11

HOPE FOR HUMAN–SPIDER RELATIONSHIPS

If you have spent any time in the United States lately, you may have heard of the giant spider invasion spreading across the eastern states. *Trichonephila clavata*, or the giant Joro spider, is native to Japan and other parts of eastern Asia. Over the past decade Joro spiders have managed to find their way into the United States and are turning out to be quite successful invaders. It's estimated that the first Joro spiders probably arrived in shipping containers around 2013. The first sightings were around the state of Georgia and they have continued spreading outwards into neighbouring states. When I first heard about Joro spiders, I jumped online looking for news articles and prepared myself for the wave of infuriating hyperbole and fearmongering. But I was pleasantly surprised. Overall, I found the articles to be much more rational and level-headed than I expected. Now I'm not going to pretend that we've completely mended spider–media relations, there were still plenty of articles leaning into the creepy-crawly factor a little too heavily. Plenty of journalists found it necessary to open with select facts, like that these were palm-sized spiders that can balloon into the skies. And there were those that found it necessary to repeatedly call them venomous spiders. While this is certainly true, given that almost every spider on the planet is venomous this is about as relevant as pointing out that they have eight legs and are made of atoms. But overall I expected much, *much* worse from the media on a story about giant Japanese spiders invading the United States. Many articles used their headlines to emphasise that these are completely harmless spiders, and even more seemed to devote space to talking about how beautiful the spiders are. And they *are* beautiful. They're golden orb-web spiders that spin enormous glistening webs, and have stunning blue, yellow and red colour patterns on their backs. When expert or public opinion was sought on the new neighbours, the talking points seemed to be about how charismatic they are, or how there might even be some benefits to having these spiders around, feeding on irritating household and agricultural pests.

As exciting as a giant Joro spider invasion can sound to a person like me, it's not necessarily something to be embraced with open arms. Any time an invasive species finds its way into new habitat there is the risk of it having an impact on the ecology of its new environment. A citizen scientist project called Joro Watch is asking people for their help in monitoring the spread of Joro spiders so that scientists can vigilantly watch for any unforeseen negative impacts of the new spider guests. One possibility that scientists are keeping an eye on is whether the new Joro spiders might outcompete native spiders, and whether native spiders need to be protected from these exotic invaders. This raises an interesting topic that many people may have never thought about – spider conservation.

Spider management and conservation

There are many spider species that have found their way out of their native habitat and into exotic new territories. With widespread global trade and shipping it's becoming much easier for small innocuous animals like spiders to hitch rides out of their native habitats and into new homes. The noble false widow, *Steatoda nobilis*, is native to the Canary Islands and Madeira but is establishing new populations across the globe. The spider is now widespread across Europe and northern Africa and in some parts of the United Kingdom is more common than native spider species. There are also reports of noble false widows in North and South America, suggesting that its spread is much larger than previously realised. The international pet trade, be it legal or illegal, is responsible for transporting many exotic species across borders. Often this can lead to accidental introductions of invasive species. For example, the Mexican red rump tarantula (*Tliltocatl vagans*) is a popular pet and, thanks to the global pet trade, can now be found in the wild in Florida.

Australia's red-back spider *Latrodectus hasselti* is making a bad name for itself by being particularly good at finding its way overseas and stirring up trouble. Both Japan and New Zealand have their own growing populations of Australian red-backs and there have been scattered reports of red-backs turning up in India, the Philippines, New Guinea and Belgium. Red-backs were found established in Osaka in 1995 and in New Zealand in the 1980s. Understandably, both Japan and New Zealand are concerned about this invasion because red-back spider bites are among the few that are venomous to humans, and this particular species absolutely loves living near people. Australian red-backs build their webs in urban spaces, particularly under outdoor furniture. In the big city of Osaka, red-backs are already present in high densities and, while they are not yet established in Japan's biggest city, Tokyo, there have been a number of sightings there and research has shown that Tokyo has the perfect climate and habitat for red-backs to live in. Similarly, New Zealand's biggest city, Auckland, could be a future hotspot for Australian red-backs.

In addition to the potential health impacts of Australian red-backs in New Zealand, there's concern that they could threaten some of New Zealand's endemic animals, including the native katipō spider, *Latrodectus katipo*. The katipō spider is under threat from habitat loss and its population is in rapid decline. There are only a few thousand katipōs left in the wild. With the introduction of Australian red-back spiders to New Zealand there is serious concern that they could directly outcompete the declining native species. Perhaps more worrying is the fact that the two species can interbreed and hybridise in the wild, making it possible for the Australian species to mess with the genetic make-up of the native population.

These two species look very similar: both are shiny and black with a bright red stripe running down their backs. So, you have a tricky situation where one *Latrodectus* species is in decline and another invasive *Latrodectus* species is on the rise. This presents a challenge for conservationists trying to relay the message that one particular red-back spider is protected but another very similar red-back spider is an exotic pest that needs to be exterminated. Biosecurity controls are in place to help avoid any further incursions of Australian red-backs into New Zealand and to stop the spread of red-backs into other areas. Meanwhile, native katipō spiders are protected by law and killing katipō spiders is a punishable offence.

Worldwide, spider conservation faces some serious challenges. Not the least of which is convincing people that spiders require protection, like any other animal. When it comes to prioritising research and conservation dollars, it's the cute and charismatic fauna that get the lion's share. It's much easier to convince someone that we should care about the fate of a fluffy koala than a snail or a grasshopper. This is true for invertebrates in general, but particularly challenging for spiders. The same goes for convincing people to care about other unpopular animals like snakes and sharks. You can see why organisations like the World Wildlife Fund choose to use a panda for their logo and not a katipō spider. Sadly, little creepy-crawly things are just as vulnerable to the impacts of human activity as any other animal and are just as, if not more, deserving of our efforts to save them. While recent reports of a 'global insect apocalypse' were a bit overblown, it's true that many invertebrate species populations are in rapid decline.

The International Union for Conservation of Nature (IUCN) sets the standard for assessing the conservation status of plants and animals. The IUCN 'Red List of Threatened Species' catalogues information on as many species as possible and makes a judgement on their vulnerability to extinction. At best a species can be ranked as 'least concern' and at worst a species can be 'critically endangered', 'extinct in the wild', or, worst of all, 'extinct'. The IUCN Red List is the go-to resource for policymakers in understanding and prioritising investment in conservation and management. There is another IUCN category that causes the most problems for invertebrate conservation – 'data deficient'. This applies to most invertebrates and is a fancy way of saying 'we don't know'. There is an unfathomable number of invertebrate species out there and we simply don't have enough information about them to assess their vulnerability. It's also generally accepted that there are many species of invertebrates going extinct without us knowing about it and before these animals have even been discovered by humans.

In a recent survey of European spiders on the Red List, researchers found that less than 100 species of spiders across the entire continent had been assessed by the IUCN. Taking into account how many species of spiders there are in Europe, this means that we know almost nothing about the conservation status of over 99.9 per cent of European spiders. Other more charismatic invertebrates fared better, with around 97 per cent of European dragonflies and butterflies having been assessed by the IUCN. On a global scale the numbers look even worse. At the time of writing the IUCN Red List includes 365 spider species worldwide as having been assessed. Meanwhile the World Spider Catalog lists 51 064 species of spiders known to science. This means we only know something about the conservation status of around 0.007 per cent of all known spider species. When it comes to spider conservation, our lack of knowledge is our biggest hurdle.

On one hand we have an aversion to spiders leading to problems with conservation, but on the other hand we have problems with some people's fascination for spiders causing conservation problems. The wildlife trade is a major global challenge for animal conservation. Many spider species are threatened by illegal poaching to serve the exotic pet industry. Big, hairy and sometimes colourful tarantulas are the most popular spider pets traded internationally. A recent investigation into the global spider trade found that around 50 per cent of all known tarantula species can be bought and sold online. Many of these are vulnerable species. For example, some of the most sought-after species are the Mexican redknee tarantulas and their close relatives. Their numbers are threatened by illegal poaching and several species are classified as either endangered or vulnerable by the Convention on International Trade in Endangered Species of Wild Fauna and Flora (CITES). Their entire genus, *Brachypelma*, is protected under Mexican law. While it's possible to obtain spiders through legitimate breeders, we know that a large number of animals in the pet trade have been taken from the wild. The ease with which a spider can now be listed and purchased anonymously online has intensified the demand for these exotic animals and the pressure put on wild populations. There is even demand for dead spiders as specimens for display. It's common to see souvenir shops filled with beautiful orb-web spiders encased in resin paperweights, or faded funnel-webs pinned out in an elaborate box-frame with a species name, usually spelled incorrectly, printed below it as if it were some prized museum specimen.

Thankfully, attitudes to wildlife conservation are changing. People are becoming more aware that conservation is not just about saving a few celebrity species, but preserving nature as a whole. This includes preserving all of the strange and unfamiliar creepy-crawlies that maintain a functioning and healthy ecosystem. It has been shown again and again that effective education programs can reduce fear and misunderstanding of unpopular animals, and this can change people's attitudes towards conservation of those species. In recent decades it has been hard to ignore the environmental education campaigns focused on the importance of pollinators such as bees, beetles and butterflies. Through these public awareness programs, the message is getting out there that animals aren't just important for their inherent value, but for the roles that they can play in maintaining healthy ecosystems. This message is now growing to include other important animals and their roles in ecosystems: decomposers, detritivores and generalist predators. Honeybees are often used as the celebrity faces of ecosystem-focused conservation campaigns because of their importance as pollinators. This gives me great hope that other small crawling creatures can be just as easily viewed as our beloved saviours.

Bees can, and have, been the source of fear campaigns and B-grade horror films. Their stings, venom and swarming behaviour could make them easy targets for our unfounded fears, but this hasn't stopped them from becoming the modern-day face of nature preservation. Bees aren't buzzing menaces, they are bright, fluffy pollinators and makers of honey, and icons for conservation. Maybe one day spiders too will be universally loved as bright, fluffy bringers of new-age pharmaceuticals, and icons for effective biosecurity protocols. We'll just have to wait and see.

A good spider is a living spider

Efforts are underway to save some spider species. The fen raft spider *Dolomedes plantarius* has been the focus of a reintroduction program that aims to save the species in Britain. These large aquatic spiders live in marshy wetlands where they hunt on the water's surface and build nests among reeds and grasses. They occur across Europe but, due to habitat loss and fragmentation, their numbers have been declining and they are now a protected species. In Britain they had been reduced to dwindling populations at three small sites. In the 1990s a long-term Species Action Plan was put together with the ambitious goal of reintroducing the spiders to wetlands, and increasing the number of fen raft spider habitats from three to 12. The plan involved breeding fen raft spiders in captivity for release into the wild. In 2010 the reintroduction process began and several thousand spiders were released into new wetland habitats in eastern England. It wasn't long before scientists started seeing new nursery webs scattered across these new wetlands; a promising sign that the introduced spiders are prospering and breeding in their new homes. There are now four newly established populations of fen raft spiders flourishing on their own and the range of the original three populations has increased tenfold since conservation interventions began. Monitoring of these populations is ongoing and work is underway to restore more wetland habitats for further introductions.

Conservation is an overwhelming challenge, and a noble one. Every spider, worm, fungus, orchid, sponge, slime mould and elm tree you see is the living ancestor of a lineage that can be traced back to the beginning of life itself. Each lineage has traced a different path, which makes them all unique, but that journey is something that all living things share, no matter how different they may be. This, I think, is what makes all living things deserving of conservation. Of course, extinction is a normal and inevitable part of this process, but the thought that humans can be directly responsible for the termination of these lineages puts great responsibility on us to make the right decisions.

And while I personally think that spiders deserve to be saved because of their inherent value as living things, there are much more pragmatic reasons for keeping spiders around. We've already talked about their potential to contribute to medicine and engineering, but we're also learning more about the practical benefits from simply having spiders living side by side with us. Just as we have come to appreciate the importance of pollinators for the ecosystem services they provide, we are recognising the importance of small predators like spiders in maintaining healthy ecosystems. If that's not pragmatic enough for you, these benefits apply to agricultural landscapes as well. Wild spiders are now recognised as an important component of what is called integrated pest management (IPM). Rather than using the misguided old-school approach of sterilising the landscape by slathering crops in pesticides and killing anything that moves, IPM advocates for using a more sensible mixed approach that exploits the ecosystem services of living things for our benefit. Since spiders are such abundant predators of small insects, their services can be used to help keep crop pests at bay. And having a diverse spider community is ideal. Imagine a crop where orb-web spiders are left to tackle flying insects in the air, while ground-foraging spiders take care of crawling insects, aquatic spiders can manage aquatic insect larvae in the nearby dam, and jumping spiders crawling along the plants can take out small crawling grubs. And there is absolutely no chance that those spiders are going to eat, burrow into, or lay eggs in whatever crop you are growing. Combine this approach with things like conservative and targeted pesticide use and effective quarantine, and that's IPM in a nutshell. And while this simple hypothetical sounds like a bit of an imaginary utopia, the evidence for spiders as biocontrol has been growing for over a century and continues to accumulate. There are a slew of studies showing that spiders can help reduce pests in everything from wheat fields in Germany, to brassicas in Australia, to grapevines in the United States. And there are other studies showing that inappropriate pesticide use can actually increase the amount of pest damage by reducing the numbers of natural predators that would otherwise have controlled the pests.

Spiders' pest control services don't stop at the farm gates, they extend into our homes and cities. Spiders are ideal urban companions that may help control pests like mosquitos. In areas where mosquito-borne diseases like malaria and dengue are prevalent, reducing disease transmission requires a multi-pronged approach, and studies are beginning to consider the role that spiders can play in controlling mosquito numbers. Spiders are ideal candidates for urban pest management since they are generally harmless, well adapted to urban habitats and pose little inconvenience to people other than the odd cobweb here and there. Unlike other household invertebrate guests, spiders won't swarm around your sugar jar, crawl into your breakfast cereal, nibble holes in your clothes, or burrow into the wooden foundations. What spiders will do is happily hunt down the other annoying insects that find their way into our houses, while mostly staying out of our way. Think about that next time you see a spider in your house minding its own business. Would you rather a small innocuous spider in the corner of the room, or a swarm of fruit flies around your fruit bowl, silverfish in your pantry, or cockroaches under the sink?

A bright future for better spider stories

This book is far from a comprehensive catalogue of every amazing spider story there is. There are endless fantastic spider stories; like how there are parasitic wasps and flies that lay their eggs inside spiders, who stay alive until the larvae hatch and then eat their way out from the inside. Or about how scientists are building miniature virtual reality environments for jumping spiders to get inside their brains and understand how they navigate. Or how, in 2020, the arachnology community was shocked by a scandal with claims of widespread data falsification, leading to the section on social spiders in this book being rather shorter than it would otherwise have been. If this book has piqued your interest in spiders then there are plenty more incredible stories out there waiting to be told.

Hopefully, these stories have convinced you beyond reasonable doubt that spiders are not the scary, alien-like creatures we sometimes imagine them to be. They are our constant companions, sharing our homes and gardens. They are the inspirational animals that bond high school classmates together for 20 years. They are the charismatic faces that advertising giants like Google and YouTube use to excite kids about space travel, because apparently astronauts and rocket ships just aren't interesting enough. They are the inspiration for new medicines and biomaterials. They are mothers, problem-solvers, weavers, astronauts and submariners.

I have spent much of this book arguing that humans and spiders can have a better relationship than we currently do. Though seeing how prolifically spider fears and misinformation seem to spread through our collective consciousness, I have wondered whether this is even possible. Having now written this book, I can confidently say yes, I do think it's possible, and signs of hope and change are all around us.

We can't underestimate the importance of Australian peacock spiders and their viral dance videos for building bridges between people and spiders. The message is getting out there that jumping spiders are captivating little fluff balls, and other spider groups can be just as charming. The more I look, the more I see cute plush jumping spider toys on store shelves. Which is a nice change from the usual creepy-crawly spider toys and Halloween decorations. For many kids now, their introduction to spiders is likely to be through the adorable *Lucas the Spider*. Lucas is a small, fluffy animated jumping spider with impossibly large eyes, who scuttles about a house and garden meeting other equally adorable critters. Much like the dancing peacock spider videos, Lucas was introduced to the world through a viral short film on YouTube, created by animator Joshua Slice. Following its sudden worldwide success, *Lucas the Spider* is now a long-running series, and has been acquired by Canadian animation studio Fresh TV, with some speculating that a *Lucas the Spider* feature film is in the works. A newly discovered Iranian jumping spider has been named in honour of Lucas, with the species name *Salticus lucasi*.

I began this book with a story about a natural history museum in Sydney having to change the entrance to its spider exhibit so it wouldn't scare away so many people. I cringed at what I thought this said about our relationship with spiders. But I have another story about something that happened around the same period. I used to pass by this museum every day on my way to work. Around this time the city became littered with billboards and posters for a new documentary series called the *Magical Land of Oz*. This three-part series focused on Australia and its incredible plants and animals. It was a big deal at the time. It was what they call a 'blue-chip' documentary, the kind that takes years and millions of dollars to capture mind-boggling footage no-one has ever seen before. We're used to seeing these kinds of documentaries narrated in David Attenborough's husky tones, but being an Australian production, this was narrated by the late great Australian icon Barry Humphries, also known by his alter ego Dame Edna Everage. A wildlife documentary of this scale and quality hadn't been produced in Australia for many years. People were thrilled to see something like this coming out of local studios that would showcase Australia in all its natural glory to the rest of the world. And when I walked through the city streets gazing at the massive advertisements for this new series, what animal do you think I saw put front and centre? A cuddly koala? A noble kangaroo? Nope. It was a spider. Having spent millions of dollars on a world-class production, the producers had decided that Australia's most glorious icon was a bright, shimmering peacock spider, dancing proudly and patriotically. Every time a bus passed by with a giant peacock spider plastered to the side of it, my heart swelled with national pride. Through this series the rest of the globe was being told that Australia was no longer the land of deadly spiders, we were the land of dancing spiders.

I have one final thing to ask of you; next time you find a spider in your house, get up close and have a good look at it. Stick a glass jar over the top and slide some paper underneath so you can turn it over in your hands and check out its shiny fangs and delicate legs. Show it to the family and see if you can find out what type of spider it is by looking it up in a field guide or online. Many local zoos and museums have outreach centres where you can send photos or bring in specimens to be identified by a local expert. Once you start appreciating the spiders around your house then you can start telling your own spider stories. Better spider stories than the old, tired stereotypes. Give them names. Then introduce your visitors to Merv the kitchen huntsman, or Dolores the garden orb-web spider. Before you know it, you'll be turning over rocks in the garden looking for more, boring your family with stories about the spider silk pharmaceutical company you've invested in, and booking a holiday to the Himalayas, not to scale a mountain peak but to catch a fleeting glimpse of the Himalayan jumping spider. Because once you start on your spider appreciation journey there really is no turning back. Spiders are awesome.

THANK YOU

This book was inspired by the work of many incredible scientists, many of whom I am lucky to call my colleagues, mentors and friends. Aaron Harmer, Anne Wignall, Anne Gaskett, Lizzy Lowe, Daiqin Li, Stuart Harris, Dinesh Rao, Jasmin Ruch, Jonas Wolff, Jay Stafstrom, Zhanqi Chen, Shichang Zhang and Felipe Gawryszewski (among many others) have made many of the discoveries mentioned in this book and have been gracious and supportive in helping me share them here. Thank you to Mariella Herberstein and Greg Holwell for supporting everything that I do and for keeping the wonder of discovery and natural history alive. Thanks, and apologies, to Matthew Bulbert and Patricio Lagos for the whole car-accident-in-the-Malaysian-jungle incident, and to Chrissie Painting for being a better jungle explorer than I ever was. Thanks to Michael Kasumovic – the screaming arachnologist in the basement – and to Jim McLean, Helen Smith and Caleb Nicholson for allowing me to use their photos.

My heartfelt thanks go to the Australian Spiders in Space team for sharing their stories and being so gracious in allowing me to retell them here, particularly Bronwyn Pratt, Greg Carstairs, Lachlan Thompson and Patrick Honan.

This book would never have been published without the encouragement of writers such as Helena Pastor, Sophie Masson, and Ian Wynne and his wonderful writers' group.

The New England Writers Centre and Varuna – the National Writers' House – graciously gave me the time, space and encouragement to turn this idea into reality. Special thanks to Deborah Bower for the late-night writing and life-coaching sessions, and to the team at NewSouth Publishing including Harriet McInerney, Sophia Oravecz, Rosie Marson, Josephine Pajor-Markus and Fiona Sim for believing in this book and giving me the opportunity to bring it to the world.

A big thank you to my mum, Marian O'Hanlon, for being very calm and collected when I jumped on planes to meet strangers from the internet promising to show me insects. And finally thank you to my wife Siobhan, and girls Mara and Edie, for giving me the life I have always dreamed of. This book is for you.

REFERENCES

This not an exhaustive list of the references that informed this book. The volume of scientific information available on spiders is vast and I imagine that a comprehensive list of every reference could be as long as the book itself. Here I have listed a number of key references for each chapter. Where possible I have pointed towards review articles that summarise fields of research, and key papers that have had a lasting impact in their respective fields. Most of the references listed here are peer-reviewed research papers published in academic journals – which means that many of them are hidden behind publisher paywalls. If you would like to read some of these dense technical papers and cannot access them freely, I recommend you email the authors of the paper, tell them how amazing they are and how much you appreciate all of the hard work they do, then kindly request a PDF of the research paper. More often than not, scientists are happy to share their work and would love to hear from their adoring fans. All species names used in the book are those listed by the online database, the World Spider Catalog, at the time of writing. I hope this gets you started on your journey down a never-ending rabbit hole of exciting spider science.

Prologue and **Why don't we like spiders?**

Muris, P., and Merckelbach, H. (1996). 'A comparison of two spider fear questionnaires'. *Journal of Behavior Therapy and Experimental Psychiatry 27*, 241–44.

Payne, R.S., and McVay, S. (1971). 'Songs of humpback whales'. *Science 173*, 585–97.

Vetter, R.S. (2013). 'Arachnophobic entomologists'. *American Entomologist 59*, 168–75.

Vetter, R.S., Draney, M.L., Brown, C.A., Trumble, J.T., Gouge, D.H., Hinkle, N.C., and Pace-Schott, E.F. (2018). 'Spider fear versus scorpion fear in undergraduate students at five American universities'. *American Entomologist 64*, 79–82.

Wignall, A.E., and Taylor, P.W. (2011). 'Assasin bug uses aggresive mimicry to lure spider prey'. *Proceedings of the Royal Society B 278*, 1427–33.

World Spider Catalog (2023). *World Spider Catalog*. Version 24. Natural History Museum Bern. <https://wsc.nmbe.ch/> accessed 8 March 2023.

Jumping spiders: The gateway to a spider addiction

Bulbert, M.W., O'Hanlon, J.C., Zappetini, S., Zhang, S., and Li, D. (2015). 'Sexually selected UV signals in the tropical ornate jumping spider, *Cosmophasis umbratica* may incur costs from predation'. *Ecology and Evolution 5*, 914–20.

Cross, F.R., Carvell, G.E., Jackson, R.R., and Grace, R.C. (2020). 'Arthropod intelligence? The case for *Portia*'. *Frontiers in Psychology 11*, 568049.

Dunn, R.A. (1957). 'The peacock spider'. *Walkabout*, April, 38–39.

Harland, D.P., and Jackson, R.R. (2000). '"Eight-legged cats" and how they see – a review of recent research on jumping spiders (Araneae: Salticidae)'. *Cimbebasia 16*, 231–40.

Jackson, R.R., and Wilcox, R.S. (1998). 'Spider-eating spiders'. *American Scientist 86*, 350–57.

Jackson, R.R., Nelson, X.J., and Sune, G.O. (2005). 'A spider that feeds indirectly on vertebrate blood by choosing female mosquitoes as prey'. *Proceedings of the National Academy of Sciences 102*, 15155–60.

Jackson, R.R., Li, D., Woon, J.R.W., Hashim, R., and Cross, F.R. (2014). 'Intricate predatory decisions by a mosquito-specialist spider from Malaysia'. *Royal Society Open Science 1*, 140131.

Maratus (2016). Ronin Films.

Otto, J.C., and Hill, D.E. (2011). 'An illustrated review of the known peacock spiders of the genus *Maratus* from Australia, with description of a new species (Araneae: Salticidae: Euophryinae)'. *Peckhamia 96*, 1–27.

Spiders in disguise

Allan, R.A., Capon, R.J., Brown, W.V., and Elgar, M.A. (2002). 'Mimicry of host cuticular hydrocarbons by Salticid spider *Cosmophasis bitaeniata* that preys on larvae of tree ants *Oecophylla smaragdina*'. *Journal of Chemical Ecology 28*, 835–48.

Angus, J. (1882). 'Protective change of colour in a spider'. *American Naturalist 16*, 1010.

Eberhard, W.G. (1980). 'The natural history and behavior of the bolas spider *Mastophora dizzydeani* sp. n. (Araneidae)'. *Psyche* (Stuttgart) *87*, 143–69.

Heiling, A.M., Herberstein, M.E., and Chittka, L. (2003). 'Crab-spiders manipulate flower signals'. *Nature 421*, 3345.

Nelson, X.J., and Jackson, R.R. (2009). 'Aggressive use of Batesian mimicry by an ant-like jumping spider'. *Biology Letters* 5, 755–57.

Painting, C.J., Nicholson, C.C., Bulbert, M.W., Norma-Rashid, Y., and Li, D. (2017). 'Nectary feeding and guarding behavior by a tropical jumping spider'. *Frontiers in Ecology and the Environment 15*, 469–70.

Reiskind, J. (1976). '*Orsima formica*: A Bornean Salticid mimicking an insect in reverse'. *Bulletin of the British Arachnological Society 3*, 235–36.

Théry, M., and Casas, J. (2009). 'The multiple disguises of spiders: Web colour and decorations, body colour and movement'. *Philosophical Transactions of the Royal Society B 364*, 471–80.

Vieira, C., Ramires, E.N., Vasconcellos-Neto, J., Poppi, R.J., and Romero, G.Q. (2017). 'Crab spider lures prey in flowerless neighborhoods'. *Scientific Reports 7*, 1–7.

Yu, L., Xu, X., Zhang, Z., Painting, C.J., Yang, X., and Li, D. (2021). 'Masquerading predators deceive prey by aggressively mimicking bird droppings in a crab spider'. *Current Zoology 68*, 325–34.

The hunter's arsenal

Bomgardner, M.M. (2017). 'Spider venom: An insecticide whose time has come? Bioinsecticide maker Vestaron says fruit and vegetable farmers are ready for its spider venom peptide'. *Chemical Engineering News 95*, 30–31.

Fernandez-Rojo, M.A., Deplazes, E., Pineda, S.S., Brust, A., Marth, T., Wilhelm, P., Martel, N., Ramm, G.A., Mancera, R.L., Alewood, P.F., et al. (2018). 'Gomesin peptides prevent proliferation and lead to the cell death of devil facial tumour disease cells'. *Cell Death Discovery 4*, 19.

Han, S.I., Astley, H.C., Maksuta, D.D., and Blackledge, T.A. (2019). 'External power amplification drives prey capture in a spider web'. *Proceedings of the National Academy of Sciences 116*, 12060–65.

Hauke, T.J., and Herzig, V. (2017). 'Dangerous arachnids: Fake news or reality?' *Toxicon 138*, 173–83.

Herberstein, M.E., Craig, C.L., Coddington, J.A., and Elgar, M.A. (2000). 'The functional significance of silk decorations of orb-web spiders: A critical review of the empirical evidence'. *Biological Reviews 75*, 649–69.

Hoese, F.J., Law, E.A.J., Rao, D., and Herberstein, M.E. (2006). 'Distinctive yellow bands on a sit-and-wait predator: Prey attractant or camouflage?' *Behaviour 143*, 763–81.

Isbister, G.K., Gray, M.R., Balit, C., Raven, R.J., Stokes, B.J., Porges, K., Tankel, A.S., Turner, E., White, J., and Fisher, M.M. (2005). 'Funnel-web spider bite: A systematic review of recorded clinical cases'. *Medical Journal of Australia 182*, 407–11.

Machkour-M'Rabet, S., Hénaut, Y., Winterton, P., and Rojo, R. (2011). 'A case of zootherapy with the tarantula *Brachypelma vagans Ausserer*, 1875 in traditional medicine of the Chol Mayan ethnic group in Mexico'. *Journal of Ethnobiology and Ethnomedicine 7*, 1–7.

Meehan, C.J., Olson, E.J., Reudnik, M.W., Kyser, T.K., and Curry, R.L. (2009). 'Herbivory in a spider through exploitation of an ant–plant mutualism'. *Current Biology 19*, R892–93.

Nyffler, M., and Knörnschild, M. (2013). 'Bat predation by spiders'. *PLoS One 8*, e58120.

Nyffler, M., and Pusey, B.J. (2014). 'Fish predation by semi-aquatic spiders: A global pattern'. *PLoS One 9*, e99459.

Peigneur, S., Elena de Lima, M., and Tytgat, J. (2018). 'Phoneutria nigriventer venom: A pharmacological treasure'. *Toxicon 151*, 96–110.

Saez, N.J. (2019). 'Versatile spider venom peptides and their medical and agricultural applications'. *Toxicon 158*, 109–26.

Silva, C.N., Nunes, K.P., Torres, F.S., Cassoli, J.S., Santos, D.M., Almeida, F.D.M., Matavel, A., Cruz, J.S., Santos-Miranda, A., Nunes, A.D.C., et al. (2015). 'PnPP-19, a synthetic and nontoxic peptide designed from a *Phoneutria nigriventer* toxin, potentiates erectile function via NO/cGMP'. *Journal of Urology 194*, 1481–90.

Stafstrom, J.A., Menda, G., Nitzany, E.I., Hebets, E.A., and Hoy, R.R. (2020). 'Ogre-faced, net-casting spiders use auditory cues to detect airborne prey'. *Current Biology 30*, 5033–39.

Suresh, P.B., and Zschokke, S. (2003). 'Webs of Theridiid spiders: Construction, structure and evolution'. *Biological Journal of the Linnean Society 78*, 293–305.

Tim, L., Herzig, V., von Reumont, B.M., and Vilcinskas, A. (2022). 'The biology and evolution of spider venoms'. *Biological Reviews 97*, 163–78.

Vetter, R.S., and Isbister, G.K. (2008). 'Medical aspects of spider bites'. *Annual Review of Entomology 53*, 409–29.

Wang, B., Yu, L., Ma, N., Zhang, Z., Gong, D., Liu, R., Li, D., and Zhang, S. (2022). 'Conspicuous cruciform silk decorations deflect avian predator attacks'. *Integrative Zoology 17*, 689–703.

Weng, J.-L., Barrantes, G., and Eberhard, W.G. (2006). 'Feeding by *Philoponella vicina* (Araneae, Uloboridae) and how Uloborid spiders lost their venom glands'. *Canadian Journal of Zoology 84*, 1752–62.

Myths and misconceptions

Isbister, G.K., and Gray, M.R. (2003). 'White-tail spider bite: A prospective study of 130 definite bites by *Lampona* species'. *Medical Journal of Australia 179*, 199–202.

Mammola, S., Malumbres-Olarte, J., Arabesky, V., Barrales-Alcalá, D.A., Barrion-Dupo, A.L., and Scott, C. (2022). 'The global spread of misinformation on spiders'. *Current Biology 32*, R871–73.

Mammola, S., Nanni, V., Pantini, P., and Isaia, M. (2020). 'Media framing of spiders may exacerbate arachnophobic sentiments'. *People and Nature 2*, 1145–57.

Vetter, R.S., and Bush, S.P. (2002). 'The diagnosis of brown recluse spider bite is overused for dermonecrotic wounds of uncertain etiology'. *Annals of Emergency Medicine 39*, 544–46.

Vetter, R.S., and Visscher, P.K. (2000). 'Oh, what a tangled web we weave: The anatomy of an internet spider hoax'. *American Entomologist 46*, 221–23.

Zobel-Thropp, P.A., Mullins, J., Kristensen, C., Kronmiller, B.A., David, C.L., Breci, L.A., and Binford, G.J. (2019).

'Not so dangerous after all? Venom composition and potency of the Pholcid (daddy long-leg) spider *Physocyclus mexicanus*'. *Frontiers in Ecology and Evolution 7*, 256.

Eight-legged bioengineers: The science of silk

Agnarsson, I., Kuntner, M., and Blackledge, T.A. (2010). 'Bioprospecting finds the toughest biological material: Extraordinary silk from a giant riverine orb spider'. *PLoS One 5*, e11234.

Blackledge, T.A. (2012). 'Spider silk: A brief review and prospectus on research linking biomechanics and ecology in draglines and orb webs'. *Journal of Arachnology 40*, 1–12.

Bon, F.X. (1710). 'A discourse upon the usefulness of the silk of spiders by Monsieur Bon, President of the Court of Accounts, Aydes and Finances, and President of the Royal Society of Science at Montpellier. Communicated by the Author'. *Philosophical Transactions of the Royal Society 27*, 2–16.

Harmer, A.M.T., Blackledge, T.A., Madin, J.S., and Herberstein, M.E. (2011). 'High-performance spider webs: Integrating biomechanics, ecology and behaviour'. *Journal of the Royal Society Interface 8*, 457–71.

Hennecke, K., Redeker, J., Kuhbier, J.W., Strauss, S., Allmeling, C., Kasper, C., Reimers, K., and Vogt, P.M. (2013). 'Bundles of spider silk, braided into sutures, resist basic cyclic tests: Potential use for flexor tendon repair'. *PLoS One 8*, 1–10.

Lazaris, A., Arcidiacono, S., Huang, Y., Zhou, J.-F., Duguay, F., Chretien, N., Welsh, E.A., Soares, J.W., and Karatzas, C.

(2002). 'Spider silk fibres spun from soluble recombinant silk produced in mammalian cells'. *Science 295*, 472–79.

Levene, D. (2012). 'Golden cape made with silk from a million spiders – in pictures'. *The Guardian*. <https://www.theguardian.com/artanddesign/gallery/2012/jan/23/golden-silk-cape-spiders-in-pictures> accessed 5 March 2023.

Lewis, R. (1996). 'Unraveling the weave of spider silk'. *BioScience 46*, 636–38.

Morgan, E. (2016). 'Sticky layers and shimmering weaves: A study of two human uses of spider silk'. *Journal of Design History 29*, 8–23.

Rising, A., Widhe, M., Johansson, J., and Hedhammar, M. (2011). 'Spider silk proteins: Recent advances in recombinant production, structure–function relationships and biomedical applications'. *Cellular and Molecular Biosciences 68*, 169–84.

Rolt, D.B. (1831). 'Silk from Spiders'. *Transactions of the Society, Instituted at London, for the Encouragement of Arts, Manufactures, and Commerce 48*, 234–36.

Scheibel, T. (2004). 'Spider silks: Recombinant synthesis, assembly, spinning, and engineering of synthetic proteins'. *Microbial Cell Factories 3*, 1–10.

'The silk-producing spider of Madagascar' (1900). *Scientific American 83*(9), 133.

Zhang, S., Piorkowski, D., Lin, W.-R., Lee, Y.-R., Liao, C.-P., Wang, P.-H., and Tso, I.-M. (2019). 'Nitrogen inaccessibility protects spider silk from bacterial growth'. *Journal of Experimental Biology 222*, jeb214981.

Birth, sex, then death

Andrade, M.C. (1996). 'Sexual selection for male sacrifice in the Australian redback spider'. *Science 271*, 70–72.

Chen, Z., Corlett, R.T., Jiao, X., Liu, S.-J., Charles-Dominique, T., Zhang, S., Li, H., Lai, R., Long, C., and Quan, R.-C. (2018). 'Prolonged milk provisioning in a jumping spider'. *Science 362*, 1052–55.

Doran, N.E., Richardson, A.M.M., and Swain, R. (2001). 'The reproductive behaviour of the Tasmanian cave spider *Hickmania troglodytes* (Araneae: Austrochilidae)'. *Journal of Zoology 253*, 405–18.

Fromhage, L., and Schneider, J.M. (2005). 'Safer sex with feeding females: Sexual conflict in a cannibalistic spider'. *Behavioral Ecology 16*, 377–82.

Ghislandi, P.G., Albo, M.J., Tuni, C., and Bilde, T. (2014). 'Evolution of deceit by worthless donations in a nuptial gift-giving spider'. *Current Zoology 60*, 43–51.

Guo, X., Selden, P.A., and Ren, D. (2021). 'Maternal care in mid-Cretaceous Lagonomegopid spiders'. *Proceedings of the Royal Society B 288*, 20211279.

Jackson, D.E. (2007). 'Social spiders'. *Current Biology 17*, R650.

Mason, L.D., Wardell-Johnson, G., and Main, B.Y. (2018). 'The longest-lived spider: Mygalomorphs dig deep, and persevere'. *Pacific Conservation Biology 24*, 203–206.

Moura, R.R., Vasconcellos-Neto, J., and de Olivera Gonzaga, M. (2017). 'Extended male care in *Manogea porracea* (Araneae: Araneidae): The exceptional case of a spider with amphisexual care'. *Animal Behaviour 123*, 1–9.

Schwartz, S.K., Wagner Jr, W.E., and Hebets, E. (2013). 'Spontaneous male death and monogyny in the dark fishing spider'. *Biology Letters 9*, 20130113.

Vollrath, F. (1986). 'Eusociality and extraordinary sex ratios in the spider *Anelosimus eximius* (Araneae: Theridiidae)'. *Behavioural Ecology and Sociobiology 18*, 283–87.

Wignall, A.E., and Herberstein, M.E. (2013). 'The influence of vibratory courtship on female mating behaviour in orb-web spiders (*Argiope keyserlingi*, Karsch 1878)'. *PLoS One 8*, e53057.

Yip, E.C., and Rayor, L.S. (2014). 'Maternal care and subsocial behaviour in spiders'. *Biological Reviews 89*, 427–49.

Masters of the Earth

Darwin, C.R. (1845) *Journal of researches into the natural history and geology of the countries visited during the voyage of H.M.S. Beagle round the world, under the Command of Capt. Fitz Roy, R.N.* 2nd edition. John Murray, London.

Glick, P.A. (1939). 'The distribution of insects, spiders, and mites in the air'. *United States Department of Agriculture*, Washington, D.C.

Gorham, P.W. (2013). 'Ballooning spiders: The case for electrostatic flight'. *arXiv*, 1309.4731.

Mammola, S., Michalik, P., Hebets, E., and Isaia, M. (2017). 'Record breaking achievements by spiders and the scientists who study them'. *PeerJ* 5, e3972.

Morley, E.L., and Robert, D. (2018). 'Electric fields elicit ballooning in spiders'. *Current Biology 28*, 2324–30.

Murray, J. (1826). *Experimental researches on the light and luminous matter of the glow-worm, the luminosity of the sea, the phenomena of the chameleon, the ascent of the spider into the atmosphere and the torpidity of the tortoise, etc.* W.R. McPhun, London.

Seymour, R.S., and Hetz, S.K. (2011). 'The diving bell and the spider: The physical gill of *Argyroneta aquatica*'. *Journal of Experimental Biology 214*, 2175–81.

Suter, R.B. (2013). 'Spider locomotion on the water surface: Biomechanics and diversity'. *Journal of Arachnology 41*, 93–101.

Thompson, J., and Trevaskis, L. (2018) 'Remote school students dig up the mystery of Maningrida's aquatic tarantulas'. ABC News <https://www.abc.net.au/news/2018-10-27/maningrida-students-research-diving-floodplain-tarantulas/10430354> accessed 10 May 2023.

Spiders in space

Hill, D.E. (2016). 'Jumping spiders in outer space (Araneae: Salticidae)'. *Peckhamia 146*, 1–7.

O'Hanlon, J. (2023). 'The Australian spider experiment on board the ill-fated space shuttle Columbia, 20 years ago'. ABC News. <https://www.abc.net.au/news/science/2023-02-01/space-shuttle-columbia-disaster-anniversary-spider-experiment/101899832> accessed 8 May 2023.

Thompson, L., and Mathers, N. (2006). '"Spiders in space": A collaboration between education and research'. *Science in School International Journal 1*, 41–45.

Witt, P.N., Scarboro, M.B., Daniels, R., Peakall, D.B., and Gause, R.L. (1977). 'Spider web-building in outer space: Evaluation of records from the Skylab spider experiment'. *Journal of Arachnology 4*, 115–24.

Zschokke, S., Countryman, S., and Cushing, P.E. (2021). 'Spiders in space: Orb-web-related behaviour in zero gravity'. *The Science of Nature 108*, 1.

Phobias and fables

Benchley, P. (2002). *Shark Trouble*. Random House, New York.

Bodkin, F. (2013). *D'harawal: Dreaming Stories*. Envirobook, Sussex Inlet, NSW.

Bouchard, S., Côté, S., St-Jacques, J., Robillard, G., and Renaud, P. (2006). 'Effectiveness of virtual reality exposure in the treatment of arachnophobia using 3D games'. *Technology and Health Care 14*, 19–27.

Coppleson, V. (1958). *Shark Attack*. Angus & Robertson, Sydney.

Davey, G.C.L. (1994). 'The "disgusting" spider: The role of disease and illness in the perpetuation of fear of spiders'. *Society & Animals 2*, 17–25.

Gerdes, A.B.M., Uhl, G., and Alpers, G.W. (2009). 'Spiders are special: Fear and disgust evoked by pictures of arthropods'. *Evolution and Human Behavior 30*, 66–73.

Hoffman, Y.S.G., Pitcho-Prelorentzos, S., Ring, L., and Ben-Ezra, M. (2019). '"Spidey can": Preliminary evidence showing arachnophobia symptom reduction due to superhero movie exposure'. *Frontiers in Psychiatry 10*, 354.

Neff, C. (2015). 'The Jaws effect: How movie narratives are used to influence policy responses to shark bites in Western Australia'. *Australian Journal of Political Science 50*, 114–27.

Olesen, J., Gustavsson, A., Svensson, M., Wittchen, H.-U., Jönsson, B., Group, C.S., and Council, E.B. (2012). 'The economic cost of brain disorders in Europe'. *European Journal of Neurology 19*, 155–62.

Wilson, W. (1951). 'The spider and the American Indian'. *Western Folklore 10*, 290–97.

Zimmer, A., Wang, N., Ibach, M.K., Fehlmann, B., Schicktanz, N.S., Bentz, D., Michael, T., Papassotiropoulos, A., and de Quervain, D.J.F. (2021). 'Effectiveness of a smartphone-based, augmented reality exposure app to reduce fear of spiders in real-life: A randomized controlled trial'. *Journal of Anxiety Disorders 82*, 102442.

Hope for human–spider relationships

Centre for Invasive Species and Ecosystem Health (2022) *Joro Watch*. <https://jorowatch.org/> accessed 8 May 2023.

Chuang, A., Deitsch, J.F., Nelsen, D.R., Sitvarin, M.I., and Coyle, D.R. (2023). 'The Jorō spider (*Trichonephila clavata*) in the southeastern U.S.: An opportunity for research and a call for reasonable journalism'. *Biological Invasions 25*, 17–26.

Knight, A.J. (2008). '"Bats, snakes and spiders, Oh my!" How aesthetic and negativistic attitudes, and other concepts predict support for species protection'. *Journal of Environmental Psychology 28*, 94–103.

Marshall, B.M., Strine, C.T., Fukushima, C.S., Cardoso, P., Orr, M.C., and Hughes, A.C. (2022). 'Searching the web builds fuller picture of arachnid trade'. *Communications Biology 5*, 448.

Michalko, R., Pekár, S., and Entling, M.H. (2019). 'An updated perspective on spiders as generalist predators in biological control'. *Oecologia 189*, 21–36.

Milano, F., Blick, T., Cardoso, P., Chatzaki, M., Fukushima, C.S., Gajdoš, P., Gibbons, A.T., Henriques, S., Macías-Hernández, N., Mammola, S., et al. (2021). 'Spider conservation in Europe: A review'. *Biological Conservation 256*, 109020.

Pérez-Miles, F. (2020). *New World Tarantulas: Taxonomy, Biogeography and Evolutionary Biology of Theraphosidae*. Springer Nature, London, Berlin, New York.

Schönfelder, M.L., and Bogner, F.X. (2017). 'Individual perception of bees: Between perceived danger and willingness to protect'. *PLoS One 12*, e0180168.

Smith, H., Clarke, D., Heaver, D., Hughes, I., Pearce-Kelly, P., and Sainsbury, T. (2013). 'Translocation and augmentation of the fen raft spider populations in the UK'. *Global Re-introduction Perspectives: 2013. Further Case Studies from Around the Globe*, IUCN/SSC Re-introduction Specialist Group & Environment Agency-ABU, 1.

Vink, C.J., Derraik, J.G.B., Phillips, C.B., and Sirvid, P.J. (2011). 'The invasive Australian redback spider, *Latrodectus hasseltii* Thorell 1870 (Araneae: Theridiidae): Current and potential distributions, and likely impacts'. *Biological Invasions 13*, 1003–19.

Vink, C.J., Sirvid, P.J., Malumbres-Olarte, J., Griffiths, J.W., Paquin, P., and Paterson, A.M. (2008). 'Species status and conservation issues of New Zealand's endemic *Latrodectus* spider species (Araneae:Theridiidae)'. *Invertebrate Systematics 22*, 589–604.

PHOTOGRAPH CREDITS

Maratus harrisi by Stuart Harris
Maratus elephans by Stuart Harris
Maratus fimbriatus by Stuart Harris
Myrmarachne smaragdina by Jim McLean
Orsima ichneumon jumping spider by Caleb Nicholson
Thomisius crab spider by James O'Hanlon
Australian wolf spider by James O'Hanlon
Argiope aemula by James O'Hanlon
Australian net-casting spider by Jim McLean
Cribellate silk of net-casting spider by Jim McLean
Orb-web spider silk cape by Cmglee – Wikimedia Commons – under the Creative Commons Attribution-Share Alike 3.0 Unported license
Nephila pilipes by James O'Hanlon
Trichonephila edulis by James O'Hanlon
Arachnura higginsi by James O'Hanlon
Huntsman mother by James O'Hanlon
Fen raft spider by Helen Smith
Orb-web spider by James O'Hanlon
Golden orb-web spider by NASA